D0948108

MANAGING
STEAM

An Engineering Guide to Commercial, Industrial, and Utility Systems

Edited by Jason Makansi

Leslie Company, New Jersey

Hemisphere Publishing Corporation, New York

Library of Congress Cataloging-in-Publication Data

Main entry under title:

Managing steam.

Bibliography: p.
Includes index.
1. Steam engineering. I. Makansi, Jason. TJ275.M28
1985 621.1 85-21894

ISBN 0-89116-462-6

Preface/Introduction

Steam is one of the most important, some might say the most important, fluid used throughout industrial, commercial and institutional facilities, and certainly one of the most common. It plays a vital role at every level—from space heating an office complex to supplying heat for converting crude oil to gasoline to driving the huge turbine generators at utility plants that provide all of us with electric power.

It takes a lot of energy to generate steam, regardless of what its used for. Since the energy crises of the 1970s, the use of steam and reuse of condensate has taken on an even more important role since steam mismanagement can be directly linked to higher fuel costs. Steam consumers large and small have been busy for several years optimizing their steam systems. Steam management, in fact, is one of the critical components of an overall energy management strategy. Engineers now purchase and install equipment to make sure that the thermal and mechanical energy contained in steam is used as efficiently as economically possible.

There are several books available on steam production and on the design, operation, and maintenance of powerplants. Similarly, there are several volumes available on fluid handling and control that include steam. What is not available in one source is a blend of this technical and engineering information specifically geared for steam management, covering production, utilization, handling, transport, system optimization, and recovery. Steam being the universal fluid that is is, this almost seems like a gross oversight. But when you consider that steam management only came into its own following the energy crises, the absence of such a volume does not seem quite as illogical.

LESLIE Co recognized this hole in the library on steam management, and decided to publish this book—an objective and generic account of the state of the art in steam management.

Practicing engineers concerned with steam and energy management will find this volume to be an invaluable working tool for solving the day to day problems encountered with steam/condensate systems, as well as a key reference when designing steam production and handling facilities. In addition, students and professors will find it a comprehensive overlook of how industry manages steam.

Information contained in the book is balanced between the conceptual and/or qualitative and the quantitative/practical, and cuts across all industries under the assumption that steam is steam regardless of who uses it and where it is used. Only its applications and economics change from one user to another.

The book begins with a description of how steam, condensate, and hot water are used in various industrial, commercial, institutional, and utility sectors, and how steam is generated and distributed. Sections on waste-heat recovery, fluidized-bed boilers, and cogeneration systems—which have become more prominent today compared to a decade ago—should be especially noted.

Following this is a chapter on boiler control theory since the boiler can be viewed as the originating and end point for all steam systems.

After the control-theory chapter, the book moves into the hardware arena as it describes in detail the many different types of valves, valve components, regulators, steam traps, and metering devices at the engineer's disposal for managing steam and condensate. In important steam-system valve and controller services, hardware is matched to specific steam-management services singled out for complexity or unusualness of the engineering involved.

Closing the book is a chapter on maintaining steam systems for optimum service and longer life. Note that the appendix offers ten nomographs for estimating steam properties and other critical values for specific situations.

LESLIE Co. gratefully acknowledges the work of Jason Makansi, who served as writer, editor, and overall project manager for the book; Jack Mesagno, who wrote the section on steam-system control theory and the operations, maintenance, and inspection services chapter; Kiyo Komoda, who directed the art and copy production; the artists, Sarah Hawn, Marlene Nocerino, and Christi Pfister; and the copy editor and proofreader, Elizabeth O'Keefe.

Contents

Section IV: HARDWARE FOR STEAM-SYSTEM CONTROL

Appendix: NOMOGRÁPHS FOR STEAM DATA

SECTION I:
INDUSTRIAL STEAM SYSTEMS

Chapter One:
How industry uses
steam

Approximately 16% of the total US energy consumption goes towards making steam for use by industry. Looking at energy consumption in another way, half of industry's energy bill is allocated to steam generation—making it one of the most important fluids to be produced, controlled, and distributed throughout an industrial plant. The top six energy-consuming industries are: refining, chemical, pulp and paper, textiles, primary and fabricated metals, and food—at least according to federal government classifications.

For example, the steel industry, though it uses steam turbines to drive blast-furnace air compressors, uses most of its energy as fuel in these furnaces. Textile plants use 90% of their energy as electric power. Compare these two industries to the pulp and paper industry in which large quantities of steam are used in paper driers and in multiple-effect evaporators that drive water out of a valuable waste stream.

Another way of characterizing industry's use of steam is the variety of unit operations that demand steam within a given plant. Textile plants use most of their steam for yarn-curing, dyeing, and drying operations and for space heating of buildings. A typical refining complex will use steam indirectly for heating, evaporation, and distillation, and directly for chemical reactions or physical and chemical separation functions, and to drive mechanical equipment such as steam turbines coupled to large pumps, compressors, fans, electric generators, and eductors that maintain vacuum for certain process operations.

In addition to the six largest energy-consuming industries there are thousands of smaller manufacturing operations that are mostly mechanical in

1

nature—the automotive industry is a good example—that use steam for heating requirements or to power machinery and heavy tools. Institutional and commercial facilities use steam for space heating or to drive steam turbine/generators that produce electric power in house. In fact, space conditioning accounts for around 50% of the energy consumed by commercial buildings. Compared to the top six industrial consumers, these steam users have comparatively light requirements. Nevertheless, when the steam consumed in these very common and basic applications is added up, the total is a sizable percentage of total steam consumption.

Most steam used by industry for process and heating needs is saturated and usually below 300 psig. Superheat is of no value and is often a hindrance because of temperature-control considerations. Pressure requirements of industrial steam systems are also modest compared to electric-power plants. In space heating systems, steam is commonly generated at the pressure needed plus pressure-drop losses in supply and distribution lines. For process systems, it is usually generated at a higher-than-needed pressure to decrease steam-piping sizes and improve the economics, but also for possible future expansion.

When industry generates electric power, its steam requirements usually include superheat and higher pressures to run turbine/generators more efficiently. In fact, all steam users are finding it economical today to generate electric power instead of purchasing it from the utility. Cogeneration, as the simultaneous production of steam and electric power is called, is discussed in more depth later in this chapter.

On the following pages is a detailed account of how industry is characterized by its uses of steam. Obviously, not all industries can be covered, nor can all the uses of steam within one industry or even one plant. This section is only intended to show the breadth and scope of steam use.

Petroleum refining industry

Steam is involved in petroleum refining right from where the oil is taken out of the ground. After crude-oil fields have reached a mature state, the oil no longer flows under its own pressure. Because there are still valuable quantities left in the well, a second fluid—usually steam or water at moderate pressures—is used to force the remainder of the oil out of the ground. The technique is called water or steam flooding. Steam has the advantage of higher temperatures and its heat content sometimes aids in overcoming the viscosity and surface tension that keep the oil from flowing.

Oil-field applications represent a very crude use of steam. Rugged steam generators usually produce steam from raw water that has been treated very little if at all. What's more, the units are once-through—that is, there is no condensate collection in most cases. Still, the amount of steam needed is considerable. Once crude oil is obtained, it must then be processed into gasoline, asphalt, diesel oil, fuel oil, and a host of other products that serve either as fuel or as intermediates for other chemical processes. Oil is first distilled; heavy components and light components are separated. Light components are further separated into methane, gasoline, and naphtha. Depending on the product mix of the particular refiner, there are many subsequent processes. In thermal and catalytic crackers, hydrocrackers, and cokers, hydrocarbon chains are decomposed to improve light-oil and gasoline yields—an increasingly important function today. Residue left after the initial distillation is further processed by visbreaking and cracking to make such products as lube oil and asphalt.

Steam functions primarily on three levels in a refinery: process heating, power generation, and as direct process input. In crude distillation at atmospheric pressure, steam is injected into the tower at around 500F to lower the process temperature and keep the crude oil from decomposing. When vacuum distillation is used, additional steam is needed to operate ejecters. Coking—a thermal-cracking process—requires even more steam compared to crude distillation. Steam is used for supplying heat to a reboiler (used to re-vaporize a condensed hydrocarbon), for driving pumps, and, in the case of fluid coking processes, for fluidizing the reactor bed.

By contrast, the alkylation, isomerization, and polymerization processes— used to produce high-octane, branched-chain hydrocarbons for aviation and motor fuel—are much larger steam consumers. Most of the steam is used for heat in the reboilers. In at least two types of alkylation processes, steam is the predominant energy requirement; in polymerization, all other forms of energy are negligible compared to steam.

In addition to the high and intermediate steam pressures needed in the unit processes, much steam is consumed at low pressure for heating crude-oil and product-storage tanks, for flushing out pipes during turnarounds and shutdowns, for heat-tracing pipelines to prevent freezing in cold weather, and a variety of other miscellaneous duties.

To meet these varied needs, most refineries generate steam at high pressure—usually around 600 psig—then cascade it through a distribution system (Figure 1) for delivery at the appropriate temperatures and pressures. There is likely to be a 600-psig steam header, a 150-psig header, and a 50-psig header serving all areas of the refinery. When enough steam at one pressure is not available, steam from the next higher pressure is delivered through a pressure-reducing-valve (PRV) station.

With such a large variety of users, the refinery's steam-supply, distribution, and condensate-collection network is very complex and difficult to maintain and optimize. Nevertheless, refiners are pursuing optimization by considering the use of steam turbines in parallel with PRVs to drive process equipment or electric generators, or recompressing low-pressure steam to an intermediate level instead of exhausting it to atmosphere.

Major sources of steam for refineries include fuel-fired boilers to generate

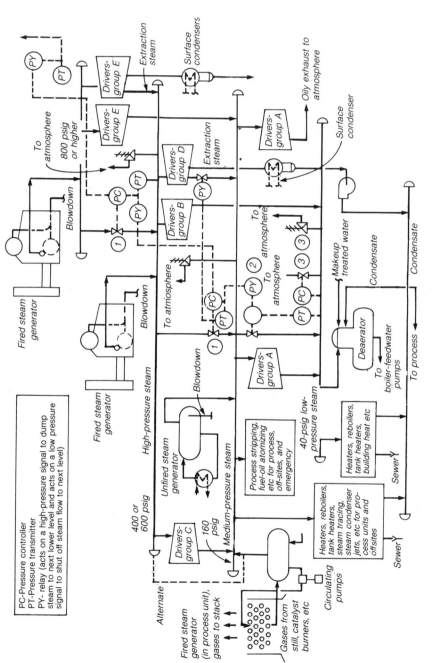

Fig 1: Refinery steam-distribution system

the highest-pressure steam, then waste-heat boilers that recover heat from furnace stacks. Stack gases from refinery furnaces are often at temperatures between 400–1000F. Waste-heat boilers compete with air preheaters in this application.

One boiler peculiar to the refining industry is the carbon monoxide (CO) boiler. Fluid-catalytic-cracker units discharge a waste-gas stream that contains about 10% by volume of CO. Energy can be recovered by burning this gas stream with supplementary fuel to generate additional steam.

The petroleum industry is one of the largest users of mechanical-drive steam turbines, mostly because they help to optimize the plant heat balance, because electric-power rates make motor drives less economical than they once were, and because steam provides more flexibility for process control at less-than-design throughputs.

Chemical industry

The chemical industry is the hardest to characterize in terms of steam consumption because its products are so diverse and specialized. It is, therefore, more illuminating to look at the various unit operations common to chemical manufacturing plants, then illustrate them with one or two examples.

Distillation is the most widely used chemical-separation technique. A mixture is brought to an elevated temperature and is separated into its components by virtue of their differences in boiling points. The vaporized material might then be condensed back to a liquid. Evaporation is a similar process where one component, the solvent, is separated from the mixture by raising the temperature to the boiling point of the solvent, sometimes at elevated pressure. Steam can be used in both evaporation and distillation to supply the heat.

Refrigeration is used to cool products or raw materials to below ambient temperature by compressing and expanding a refrigerant like ammonia or fluorocarbons. Here, steam is used in a steam turbine that drives the compressor. In a like manner, steam-turbine drives are used to pump material or to mix and blend different materials—an energy-intensive step if the material is highly viscous.

As in refining, chemical plants use steam directly in the process and for holding a vacuum on a unit operation. One source believes that this latter use of steam is obsolete today because electric motors or gas-turbine drives are more energy efficient—especially for high-flow, high-pressure applications.

Consider the production of ethylene—an important intermediate in the making of plastics, rubber, and synthetic fibers. Hydrocarbon feedstock and steam are reacted together at about 1800F. The product mixture is cryogenically cooled, then compressed to 450–600 psig after which the ethylene is distilled from its feedstock and byproduct material. Cooling, compression, and high-temperature/high-pressure reaction with steam make ethylene production a large consumer of steam in the chemical industry.

On the other hand, some chemical reactions are exothermic—that is, they produce more heat than they consume. Sulfuric acid plants are a good example of exothermic plans. Here, high-pressure steam can be produced from the reaction heat to drive a steam turbine for motive power or for electric-power generation, or the steam can be directed back to the process.

Pulp and paper industry

Paper production involves four basic steps: pulpmaking, forming, pressure, and drying. Of these, pulping and drying consume the most steam. In chemical pulping, woodchips are mixed with chemicals and cooked in digesters under steam pressure and controlled conditions to dissolve the lignin and release the fibers. Once separated, the lignin/chemical-containing waste steam is processed in multiple-effect evaporators to concentrate it. Once evaporated, the waste has a high-Btu value and is burned in special boilers called black-liquor-recovery boilers to generate additional high-pressure steam. Low-pressure steam at 35–80 psig is used in the multiple-effect evaporation step in varying quantities depending on the type of process.

Digesters are either batch or continuous. Batch processors are older and less thermally efficient. In some cases, the feed is heated via direct steam injection. Continuous digesters use indirect heating with exchangers. One source claims that continuous digesters use about 40% less steam than batch units, and consume the steam more uniformly.

After the fibers are formed into a web on the paper machine—the Fourdrinier machine is the most famous—and pressed to remove water, the sheets are dried by low-pressure steam that condenses inside iron cylinders. This is the largest steam consumer using nominally two pounds of steam for every pound of paper. Pulp bleaching also consumes steam. Bleaching is required in varying amounts according to the type of pulp and source of wood. The steam is used to control the reaction temperature in several of the tanks.

As mentioned earlier, pulp and paper plants generate much steam in recovery boilers. In fact, larger integrated paper mills in the 500–1000-ton/day capacity were cogenerating long before it became a born-again answer for industry's energy problems. As an industry, it also makes good use of extraction and backpressure steam turbines to balance steam and electric-power requirements. Smaller, non-integrated mills usually purchase most of their electric power and generate steam at the low pressure and temperatures required of the process.

Waste-fired boilers used at paper mills involve special designs that must account for the varying Btu value of the waste fuel, the inorganic chemical species in the fuel, the relatively large water content even after drying, the

corrosive nature of the waste products, and other adverse combustion characteristics. In general, they are capable of producing high-pressure/high-
temperature steam required for steam-turbine/generators to optimize
and/or maximize electric-power production. In fact, the pulp and paper
industry is in the vanguard of a movement in industry towards boilers that
operate above 1500 psig. Boiler operation above the 600–900-psig range
requires more complicated and more expensive feedwater treatment.

Food industry

Heat processes in the food industry are used to effect chemical and biolog-
ical changes to the food. The most common are indirect heating where
steam—usually saturated up to temperatures of 400F—hot water, or hot air
is used in heat exchangers without contacting the material to be heated. For
strictly process needs, steam is rarely superheated. Examples of steam-
consuming unit operations are blanching, dehydration, sterilization, pasteurization, and evaporation.

Blanching is used to prepare vegetables and fruits for canning, freezing, or
dehydration. It is intended to deactivate enzymes or to destroy enzyme
substrates such as peroxides. In immersion blanching, food is passed at a
controlled rate through a perforated tank rotating in a second tank of hot
water controlled at around 180F. Alternately, the food is suspended in
water, heated to temperature, then pumped through a holding tube.

Steam blanchers, on the other hand, use saturated steam at pressures less
than 10 psig. Food is conveyed through a steam chamber on a mesh belt or
by means of a screw conveyor.

Sterilization and pasteurization are heat-applying processes used to kill
microorganisms, spores, and/or pathogenic organisms. Heating in saturated
steam is the most commonly used method of sterilizing food in containers.
Nonagitating and agitating batch sterilizer vessels called retorts are still used
extensively although they are somewhat outdated. Steam is distributed at the
bottom of the vessel while containers are located at the top or the side
depending on the vessel configuration.

Continuous sterilizers (Figure 2) are more appropriate for today's plants.
In one type of unit, called a hydrostatic sterilizer, containers are transported
through a steam tower at around 250F, pressurized by means of a hot-water
column. These units use steam and hot water more efficiently than batch
sterilizers. Where flame sterilization is used, containers are still preheated in
a steam chamber, then further heat treated in a steam-filled holding
chamber. This method is limited to small cans.

Food sterilization can also be accomplished outside the container. For
example, dairy products and other foods can be sterilized by the Dole
process—one food process that uses superheated steam. Milk is sterilized by
injecting steam into the milk or by indirect plate heating.

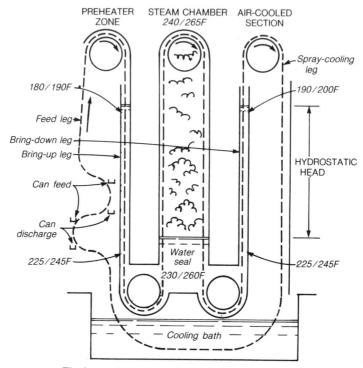

Fig 2: Continuous sterilizer for the food industry

Pasteurization proceeds at lower temperatures than sterilization—usually less than 212F. Normally, it is accomplished as a batch process where bulk food is stirred in vessels jacketed with steam or hot water, or continuously by immersing containers in hot water, by spraying hot water onto the cans, or by exposing the cans to steam at atmospheric pressure.

Evaporation is a major unit operation in the food industry that uses saturated steam to supply the sensible and latent heat of evaporation to the feed.

Inexpensive open-pan evaporators are fitted with an outer jacket filled with steam, or with internal coils filled with steam. Horizontal and vertical short-tube evaporators are more standard pieces of equipment in today's plants. They are basically shell and tube heat exchangers. In the horizontal units, the steam flows inside the horizontal tubes; the material is on the outside. A disengagement space is provided to separate liquid and vapor. In the vertical unit, steam condenses on the outside of tubes mounted vertically inside a steam chest. Options that are included on evaporators include mounting the steam tube bundles outside of the disengagement space (Figure 3).

Other common evaporators include: long-tube units, also shell and tube exchangers in which steam is usually condensing inside the shell; plate eva-

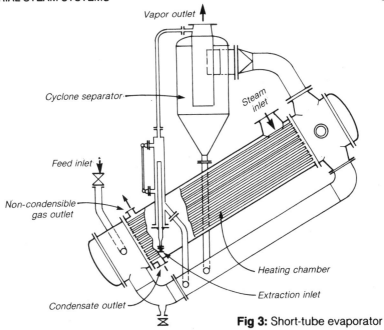

Fig 3: Short-tube evaporator

porators which employ steam only in the first of several stages of the plates; expanding-flow evaporators, centrifugal evaporators, and low-temperature evaporators that use heat pumps available for more specialized applications.

Multiple-effect evaporation is used to improve the thermal efficiency of the evaporation and to reduce steam consumption. The more effects or stages, the better the steam economy—at least until capital costs for the extra stages no longer balance the reduced costs of generating the additional steam.

Textile industry

Most of the energy used in textile mills is electric power for machinery and for air-conditioning systems. For example, spinning and texturizing processes use steam mostly for space heating. In the weaving process, steam is used for winter heating and for what's called the slashing process—part of preparing the material for weaving.

Later operations—such as woven- and knitted-fabric finishing—are a different story. Over 60% of the energy is needed as process steam for preparation, dyeing, and finishing operations. Steam is used to heat the dyeing solution prior to its being applied to the fabric, to make hot water that is later used in finishing, and to cure and heat-set such products as carpet yarn. For these needs, the steam is usually obtained from packaged oil- and gas-fired boilers that normally generate steam near the low pressure levels required.

To remove the shackles of oil and gas supply problems, some textile firms are building central coal- or waste-fired steam-generating plants to serve several plants at once. As an industry, it is in the forefront of this community energy-supply concept.

Other industries

Beyond the most energy-intensive industries, there are a variety of allied industries that use steam in many different ways. The pharmaceutical, plastics, and rubber industries, for example, are allied to the chemical industry. Consumer products such as shampoos, perfumes, soaps, and other common household goods are manufactured similarly to chemical and food-type operations. Unit operations that use steam—distillation, evaporation, drying, curing, dehydration, heating, sterilization, chemical reactions, and mechanical drives—are common to many of them.

Commercial/institutional facilities

Virtually all commercial buildings, apartment houses, office complexes, airports, factories, universities, hospitals, and the like are equipped with a heating system and many of them use steam for this purpose (Figure 4). Piping connects the source of steam to a variety of heat-transfer units such as radiators and absorption/refrigeration units. Pressure of the steam is often controlled by the outside ambient-air temperature to between subatmospheric to 80 psig and higher in most cases, with corresponding temperatures that range from 100–350F. Often, the steam-heat system must work together with the cooling system to optimize space conditioning.

Such heating systems are classified in a number of ways: by their piping configuration, by steam conditions, or by the means of causing steam and water to flow.

In a one-pipe system, a single main supplies steam to the heating unit and brings the condensate back to the steam generator. By extension, a two-pipe system provides separate steam-supply/condensate-return piping networks. It is further labelled upfeed or downfeed depending on which way the steam flows in the supply mains, and as dry return or wet return depending on whether condensate mains are above or below the water line of the condensate tank.

Alternately, steam-heating systems are designated as high pressure, if steam pressures go higher than 15 psig, low pressure when steam is between 0–15 psig, and a vacuum system if operated below atmospheric pressure. Combinations are possible as well. Likewise, for water (high-pressure condensate) systems, low temperature means up to 250F, medium temperature means 250–300F, and high temperature from 300–500F and above.

Finally, if condensate is returned solely by gravity forces, the system is

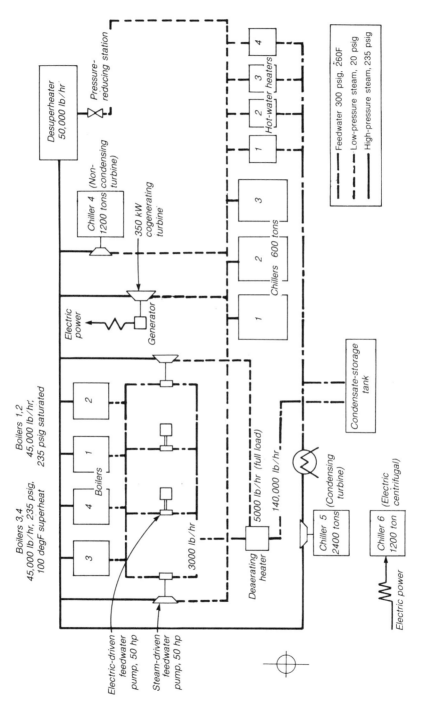

Fig 4: Commercial steam-heating system

termed a gravity feed one. When steam traps and pumps are used, it is an open or closed mechanical-return system depending on whether a vacuum pump is used.

Steam-heating systems offer these advantages over hot-water systems: 1) Optimum flexibility for overcoming heat losses, balancing heat requirements, and reducing space-temperature fluctuations, 2) smaller overall system size—supply lines are larger but return lines are smaller reducing overall size—and total weight of piping, 3) lower building costs since floors do not have to support contents of expansion tanks and additional piping weight, 4) lower pumping costs and lower maintenance costs, and 5) faster response to changes in demand.

When comfort heating is the sole objective, the steam is usually generated at the low pressures desired in the heat-transfer units. However, in most industrial facilities, the dual requirements of higher-pressure process steam and low-pressure space-heating steam dictate the need to generate steam at the higher pressure, then reduce it in pressure-reducing valves for each individual user. Distance between the steam source and users also dictates the supply pressure to overcome distribution-line losses. Finally, there is an economic imperative to generate steam at higher-than-needed pressure to take advantage of the smaller piping diameters and associated costs.

District-heating systems

An alternative to a central heating plant is a district-heating system. These can be used for clusters of buildings such as universities or office parks, and even serve entire downtown centers of large metropolitan areas.

The technique was popular in the US at the turn of the century, but stagnated after World War Two. To illustrate: In the early 1900s, there were over 150 district-heating systems in the US. After World War Two, only a handful survived because they could not compete with cheap fossil fuels and correspondingly cheap electric power. Today, many energy specialists forecast a revival of district heating in the US because it combines well with cogeneration and significantly increases overall fuel-use efficiency. Because steam-heated systems dominated over hot-water systems in the past, they are expected to do likewise during the revival. Still, hot-water systems are especially popular when the utility is not involved, such as in shopping centers, airports, etc.

In Europe, the scenario is different. Growth of district heating got started after World War Two and has continued to this day. West Germany alone has close to 500 systems. The European method employs hot water from the local utility's backpressure turbine/generators. Popularity of the systems overseas has as much to do with different electric-utility characteristics as it does with fossil fuels that have always been expensive.

Recent studies suggest that district heating is an attractive way for US utilities to serve urban areas, diversify, and boost revenues in the face of low load growth predicted for the next decade or so.

District-heating systems are similar to the central heating plants discussed earlier as far as steam generation and distribution are concerned, except that the distribution network is more complex and thus makes up 50–75% of the total capital costs. Individual users may use the steam directly for heating, reduce its pressure for a low-pressure steam-heat system, or direct it through heat exchangers to supply the heat for a hot-water system.

Chapter two:
Generating steam
in boilers

The cornerstone of any steam system is the boiler or steam generator. In its crudest sense, what's going on in a boiler is nothing more than a pot of water boiling on the stove. As a matter of fact, the first boilers on the industrial scene hundreds of years ago were quite similar to large kettles.

Inside, heat is transferred to the treated feedwater/condensate stream which raises the temperature of the water up to its boiling point. Continued application of heat vaporizes the water into steam. Once the water is converted, more heat is applied until the steam reaches the desired temperature and pressure (Figure 5). Fuel is burned in a furnace coupled to the boiler to provide the heat. Oil, natural gas, and coal are the traditional sources of heat, but today process-waste streams such as gas-turbine and diesel exhaust, and combustibles' waste products—sawdust, tires, low-Btu gas from chemical and refining operations, landfill gas, municipal solid waste and many others—are being used more and more to offset energy costs at industrial and commercial facilities. Even electric power is used to generate steam if rates are low enough.

Keep in mind that the term "boiler" is sometimes used to refer to hot-water-producing systems as well. Especially in institutional, commercial, and smaller industrial facilities, steam generation is not required and hot water suffices.

Naturally, boilers today come in all shapes, sizes, and configurations depending on a variety of factors. Some of the most important are: the type of fuel being burned in the furnace, the amount of heat in the waste stream (for waste-heat boilers), the amount of steam that the boiler must produce, the final temperature and pressure of the steam, the overall thermal effi-

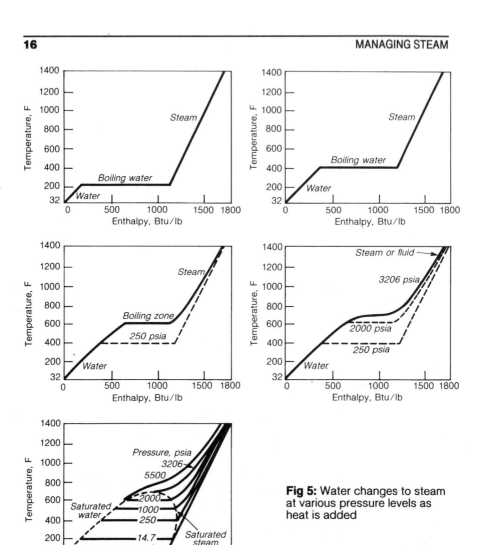

Fig 5: Water changes to steam at various pressure levels as heat is added

ciency of the heat-transfer process, the installation space available, and required pollution-control facilities.

Boilers and steam generators can be classified in a number of ways. The most common delineation, watertube and firetube boilers, describes the internals. Boilers are also classified by the heat source: thus, oil/gas-fired boilers, coal-fired boilers which are subclassified as pulverized-coal-fired and stoker-fired, solid-fuel-fired boilers, which imply solid fuels other than coal (such as woodwaste and biomass), electric boilers, and waste-heat-fired boilers implying no fuel but a high-temperature gas stream. Finally, boilers are distinguished by erection procedure. Packaged boilers are constructed assembly-line fashion, are mounted on a skid, and transported to the site in one package ready for hookup to the auxiliary piping. Shop-assembled boilers are built up of largely modular components and shipped to the site in one piece. Field-erected boilers are put together at the site after largely modular components have been fabricated and shipped to the site.

In recent years, a new type of boiler, the fluidized-bed steam generator, has emerged as an important contender for industrial steam systems. Following the energy and environmental shakeups of the 1970s, manufacturers, with much funding assistance from the government, developed the fluidized-bed boiler because it is an inherently cleaner combustion process, it can be adapted more readily and inexpensively to control major pollutants, and it can burn a variety of liquid, gaseous, and solid fuels alone or in combination.

Firetube boilers

Implied in the name firetube is that the heat is transferred from hot combustion gases flowing inside the tubes to the water surrounding them. They are sometimes referred to as shell boilers because the water and steam are contained within a single shell housing the steam-producing elements. Most firetube boilers, just as often called scotch-marine boilers, are designed to burn liquid and gaseous fuels. When designed for solid-fuel firing, they are commonly called firebox units (Figure 6).

Fig 6: Firebox boiler

Scotch-marine boilers have not changed much over the last 25 years. Combustion takes place in a cylindrical furnace, usually corrugated to increase strength, located inside a cylindrical pressure vessel. Water in the welded steel boiler is heated by firetubes that carry the combustion gases. The gases may make two, three, or four passes through the unit (Figure 7).

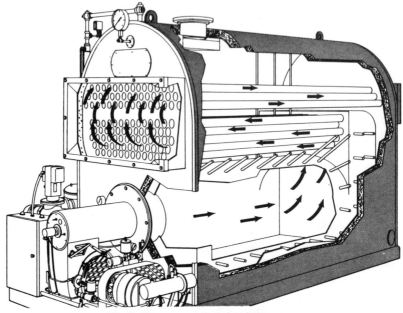

Fig 7: Three-pass firebox boiler

Increasing the number of passes increases the efficiency but there is an economic tradeoff in the need for greater fan power to move the combustion gases, increased mechanical complexity, and larger shell diameter. Four passes tends to be the economic limit.

Saturated steam up to 250 psig is near the practical limit for firetubes. Shell strength of the cylinder required to resist rupture is proportional to pressure times the diameter. High pressures and large diameters lead to prohibitively thick shells and high costs. Today, they are usually less expensive than watertube boilers without economizers up to about 40,000-lb/hr steam capacity. Other advantages of firetubes over watertubes include: they handle load changes rapidly because there's a lot of water at the saturation temperature (the disadvantage here is there is also a lot more water to heat up to operating pressure after a surge in demand), they are generally shorter than watertube units and thus requiring less headroom, they exhibit higher efficiency in many cases because vendors put about 40% more surface area into them and this permits higher heat-transfer rates. Fuel-to-steam efficiencies go as high as 86% if distillate oil is burned in special burners. Economizers, designed especially for firetube units, can boost this number even further.

Firebox boilers offer a greater ratio of furnace volume to heat-transfer surface so they can handle solid fuels with lower heat-release rates. Combustion gases normally make three passes. Superheating is not available with the firetube or the firebox design.

Watertube boilers

The watertube boiler is the backbone of American industry. They were introduced 200 years ago, but the original one looks nothing like one today. Horizontal tubes with vertical or slightly inclined sectional headers and one main steam drum, popular in the early 1900s, gave way to bent-tube designs with several drums that serve as convenient collecting points to separate the steam from the water. As boilers grew in size during the middle of the century, water-cooled furnaces were developed which eliminated the need for some of the drums.

Today, the design of watertube boilers can be characterized by the following points. In general, designers maximize the use of vertical or near vertical waterwall tubes in the radiant section (surrounding the furnace) and the convection section. Another trend is making the most of steam/water natural circulation to increase heat absorption. Finally, economizers are becoming a regular feature for energy conservation. Here, feedwater to the main steam-generator drum is heated by the flue gas after it passes by the convection section.

Most packaged watertube boilers follow structural configurations known as the A type, the D type, and the O type (Figure 8). The A type has two smaller lower drums or headers and a larger upper drum for steam and water separation. Most of the steam production occurs in the center furnace-wall tubes entering the drum.

The D-type boiler has only two drums and is considered the most flexible of the three. The more active steam-producing risers enter the drum near the water line. It features a larger combustion-chamber volume and it is the easiest to fit with a superheater or economizer.

In the O-type boiler, also a two-drum unit, the symmetry of the design exposes the least tube surface to radiant heat.

Today, packaged boilers of the above configurations up to about 250,000-lb/hr steam capacity can be shipped via railroad; larger ones, up to 600,000 lb/hr, can be fabricated but must be shipped by barge or freighter. Majority of units are designed for 125–1000-psig operating pressures even though 2000 psig is not unrealistic. Temperatures range from 350–950F.

Water carryover into the steam has been a recurring problem with packaged boilers. Steam separators are needed in all but the smallest units because there's about 20–30 lbs of water per lb of steam flowing through the drum in a typical boiler.

Several separator designs are available today. For low-pressure saturated-steam units, baffle and chevron types suffice. Centrifugal separators are used for high-pressure superheated-steam applications.

Superheaters for packaged boilers generally are the radiant or radiant/convective type. The former tends to heat steam to higher-than-design temperatures at low loads; the latter maintains a relatively constant

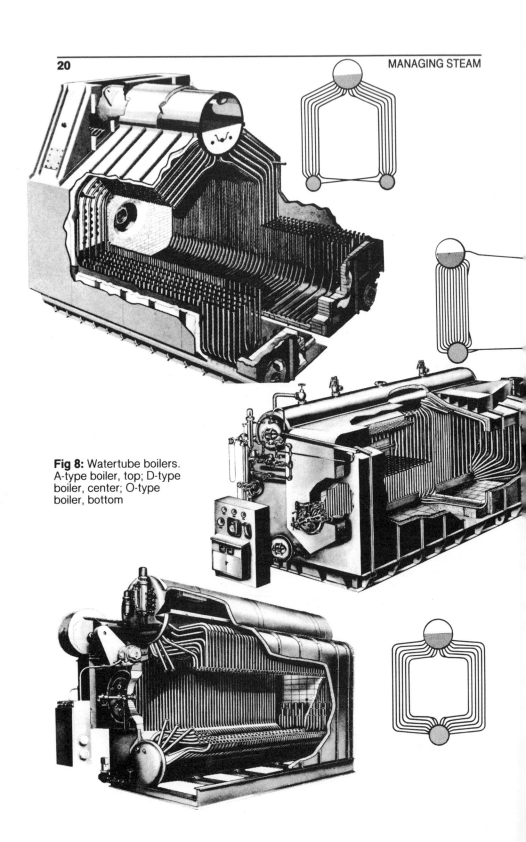

Fig 8: Watertube boilers.
A-type boiler, top; D-type
boiler, center; O-type
boiler, bottom

steam temperature over the load range. Most are of all-welded construction to eliminate tube-joint leakage.

Economizers are used to improve the efficiency of watertube boilers. They are relatively simple heat exchangers. Some units for retrofit to existing boilers have finned tubes running continuously from the inlet header to an outlet header with terminals rolled or welded. Note that finned-tube economizers are not recommended for dirty fuels, especially waste-fuel firing.

Packaged boilers designed for burning premium fuels generally lack fuel flexibility—an all-important criterion for industrial steam users—especially for burning solid fuels such as coal and wood. Reasons are that furnace heat-release rates are too high, flue-gas velocities through tube banks are too great, and tube spacings are too tight. Solid fuels are more difficult to handle in general and require a more conservative furnace/boiler design.

Some manufacturers offer shop-assembled coal-fired boilers that are com-

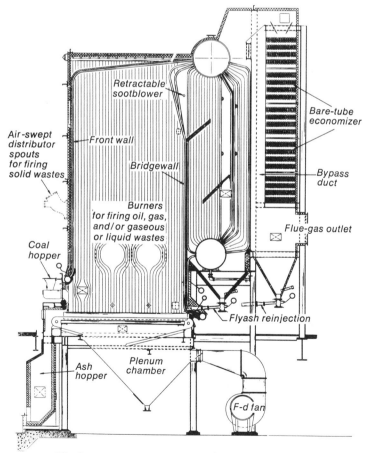

Fig 9: Field-erected boiler for coal and other fuels

patible with stoker-firing and pulverized-coal firing. However, most of the important design features involve the combustion process and will not be discussed here.

Beyond the 100,000-lb/hr size, the industrial steam user must choose among field-erected designs (Figure 9) in order to obtain good fuel flexibility. Large boilers such as these are rarely designed for only oil and gas anymore. Instead they are designed to be coal-fired systems that can be operated on oil and gas, or even a variety of waste fuels as backups.

Choosing between stoker-fired and pulverized-coal-fired units involves many engineering and economic factors. Stokers in general are limited to boilers smaller than 250,000 lb/hr—though larger units are operating. There is no theoretical upper limit to the capacity of pulverized-coal-fired units, and their fuel-to-steam efficiency is usually higher than stoker units. Pulverized-coal units also exhibit better fuel flexibility relative to oil and gas in the larger sizes, but stoker units handle a wider variety of solid-waste fuels with minimal preparation prior to combustion.

Waste-heat boilers

Producing steam is often a sound way to recover waste heat, reduce fuel consumption, and improve the overall thermal efficiency of an industrial plant. If the waste stream is above 600F, then generating steam—as opposed to heating combustion air, feedwater, or process liquids—is usually the most economical recovery method. One reason for this is that water has a greater heat capacity than other fluids so heat-transfer equipment can be made proportionately smaller. Also, the equipment can be constructed of carbon steel pressure parts instead of more expensive alloys because the evaporation process keeps the tube surfaces relatively cool even at high steam pressures and temperatures.

Choice of boiler will depend on the characteristics of the waste gas stream and the steam requirements. Mass flow of the waste stream, average temperature, pressure and specific heat of the stream, fouling, erosion, and corrosion potential of the waste material, temperature of the steam output, feedwater temperature at the boiler inlet, allowable gasside pressure drop, and temperature of the waste gas stream leaving the boiler must all be specified before the proper boiler can be selected.

Both watertube and firetube waste-heat boilers are available, and are divided into high-pressure (3 psig and above) and low-pressure gas (less than 3 psig). The former are mostly associated with the chemical and refining processes where they are used in the process stream to cool a gas stream. In this application, steam is considered a byproduct.

Firetube heat-recovery boilers are capable of generating saturated steam at pressures up to 250 psig. They come in one-, two-, and three-pass designs with or without supplementary firing, and with or without an induced-draft fan depending on whether the gas-stream pressure is sufficient to move it through the tubes. In one type of supplementary-fired unit, one set of firetubes extracts heat from the combustion gases, another the heat from the waste gases. The burner only operates when the steam demand cannot be met with heat recovery only. These are rated up to about 35,000 lb/hr for gasside temperatures up to 2000F.

Though firetubes are limited in capacity, they do offer one potential advantage over watertube units: they handle gas streams with high fouling potential.

Watertube heat-recovery units can be built for most steam capacity, pressure, and temperature, and their efficiency is usually higher than firetubes.

Heat-recovery boilers, like fuel-fired ones, are designed for either natural or forced circulation. Natural circulation is used in industry for the most part for steam pressure below 2500 psig. The basic unit consists of two drums, risers, and downcomers to produce saturated steam. Forced circulation is recommended when the boiler has to be arranged for upflow or downflow of the hot gases. Amount of heat-transfer surface required in each case is the same because the steamside coefficient is unchanged whether the fluid is forced through the tube or flows by itself. Tube surfaces can be bare or finned depending on the percentage of troublesome solids in the waste gas and the gasside temperature.

Superheaters having one or more rows of bare or finned tubes are located ahead of the generating bank where the gas temperature is hottest. Condensate-spray nozzles are sometimes used to control superheater-outlet temperature. Economizers are usually economical today for waste-heat boilers.

Gas-turbine heat-recovery steam generators (Figure 10) are an important category of waste-heat boilers. The arrangement of superheater, high-pressure boiler, economizer, and low-pressure boiler permits efficient generation of steam that often is suitable for driving steam-turbine generators. When configured this way, the system is called a combined-cycle if the gas turbine is driving a generator as well.

One of the important design parameters for gas-turbine heat recovery is the pressure drop through the boiler. It affects the turbine's backpressure and thus its output and efficiency. Turbine power output can fall by 0.3% for every 1.0-inch H_2O rise in backpressure. A second parameter is the pinch point—the difference between the temperature of the waste gas at the boiler outlet and that of the saturated steam. While it is desirable to maximize energy recovery by keeping the pinch point as low as possible, this is not practical because required heat-transfer surface increases rapidly. Finally, the economizer approach temperature is also important. Temperature of the water at the economizer outlet should be held as close as possible to that of the saturated water in the steam drum to maximize efficiency, but not too close or else boiling and steam hammer will occur.

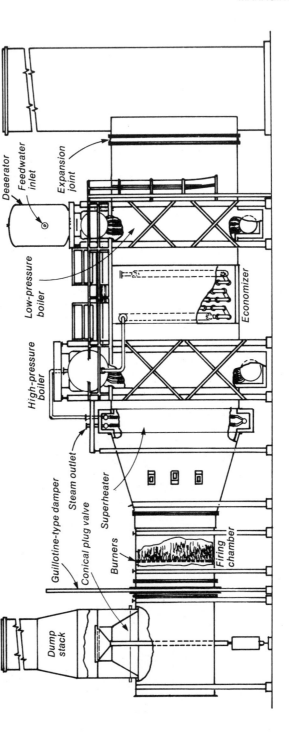

Fig 10: Gas-turbine heat-recovery boiler

Electric boilers

Electric boilers and hot-water generators were introduced in the US more than 50 years ago. But it was not until the early 1970s that they became economically attractive for producing steam in industrial plants. When fuel-oil prices and environmental regulations hit, the electric boiler was a quick, non-polluting way to generate additional steam.

Even today, in the face of rising electric-power costs, electric boilers find uses in leveling out load demands and taking advantage of off-peak charges from the utility. They can be used as standby or startup boilers replacing fossil-fired boilers, which are at least 25% more in capital costs. Because these boilers operate only a few hours out of the year, their high operating costs do not plague the owner. Finally, they can supply backup steam or essential steam to industrial and central stations in the event that the less expensive source is interrupted.

The two basic types of electric boilers are the resistance and the electrode. In resistance units, current is passed through a series of resistance-type heating elements—high-resistance wire encased in an insulated metal sheath—which are submerged in water. In electrode units (Figure 11), the

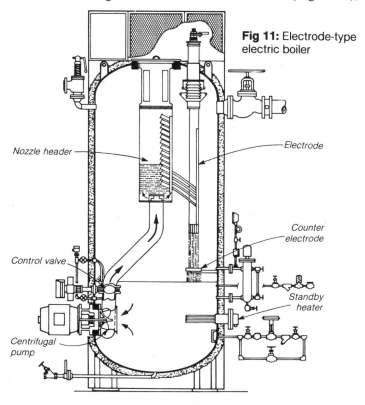

Fig 11: Electrode-type electric boiler

Nozzle header

Electrode

Counter electrode

Control valve

Standby heater

Centrifugal pump

current is passed through the water and not the wire; here, the liquid itself converts the electrical energy into hot water or steam. Industrial application of the resistance units is economical only up to sizes of 10,000 lb/hr of saturated steam. Electrode boilers are offered as either low-voltage or high-voltage units. Steam-pressure capability extends up to 500 psig; higher pressures are possible but there is an inherent risk of insulator breakdown, and pressure-vessel costs go up fast.

Most electric boilers are used in conjunction with oil- and gas-fired boilers. In a typical arrangement, a demand controller allows the electric boiler all the power available at favorable rates and the steam produced is sent to a header at a slightly higher pressure than that from the fired boiler. If header pressure drops below the demand, the combustion control system on the conventional steam generator responds and that unit picks up the slack.

Key to successful electric-boiler operation is water treatment. They are less forgiving than fossil-fired units under changes in water-system chemistry.

Fluidized-bed boilers

In its simplest form, the fluidized-bed boiler is a new type of combustor/furnace coupled to a conventional boiler heat-transfer surface. Solid, liquid, or gaseous fuels together with an inert material such as sand, silica, ash from the combustion process, and/or limestone if SO_2 removal is required, are kept in suspension in the furnace through the action of fluid-

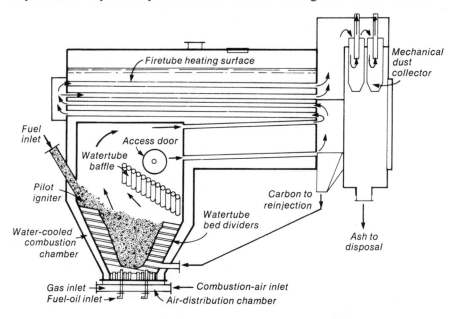

Fig 12: Firetube fluidized-bed boiler

izing/combustion air distributed from below. The turbulence and better mixing, compared to a conventional burner, provide a more uniformly distributed temperature and a much lower combustion temperature. This in turn allows a smaller furnace volume for equivalent heat release and lower NO_x formation. Steam-generating tubes or superheating tubes are sometimes immersed in the bubbling bed and, in some units, waterwalls enclose the furnace. Above the bed, the hot combustion gases are cooled by boiler-tube convection passes in the freeboard area.

Fluidized-bed boilers also come in watertube and firetube designs (Figure 12). One popular firetube design has a vertical combustion chamber combined with horizontal firetubes. Feedwater is heated in three areas: in tubes immersed in the bed, in tubes placed laterally in the freeboard zone above the bed, and by firetube surface immersed in the water. A tapered combustion chamber extending downward forms a full water-cooled chamber for the bed.

At the present state of the art, the conventional bubbling-bed boiler suffers from lower combustion efficiency compared to a stoker-fired, pulverized-coal-fired, or other solid-fuel-fired boiler. One way designers are increasing efficiency is augmenting the bubbling bed with a solids-reinjection system. Combustible material that elutriates (carried out with the combustion gases) from the bed is directed back to the furnace, giving the solids more residence time for complete burnout. Another advantage of reinjection systems is being able to use coarser particles in the bed with less chance of their escaping.

Circulating beds (Figure 13) take reinjection to the extreme. Solids carryover is actually promoted with high air velocities through the bed and the solid material is continually recirculated. The technique increases residence time, enhancing complete burnout, and if limestone is used, providing more time for the SO_2-absorption reactions to take place.

Some of the early fluidized-bed units displayed unacceptable erosion and corrosion rates of the tubes immersed in the solids. One way out of this problem was to separate the heat-transfer tubes from the bed, but the heat had to be absorbed somewhere to control bed temperature. Several designers now incorporate in-bed circulation, which increases heat transfer to the water-cooled walls. This concept is seen mostly in the smaller watertube and firetube units popular with industry.

In one circulating system, the entire shell including the solids riser is enclosed by waterwalls. There are no steam-generating tubes in the bed. Instead, superheating and steam generating take place in a convection section above the bed. Some designs separate out the heat-transfer and solids-separation sections, but all of them have water-cooled combustion chambers.

Circulating systems have better load-following capability than conventional bubbling beds, because of the separation of the solids heat exchange

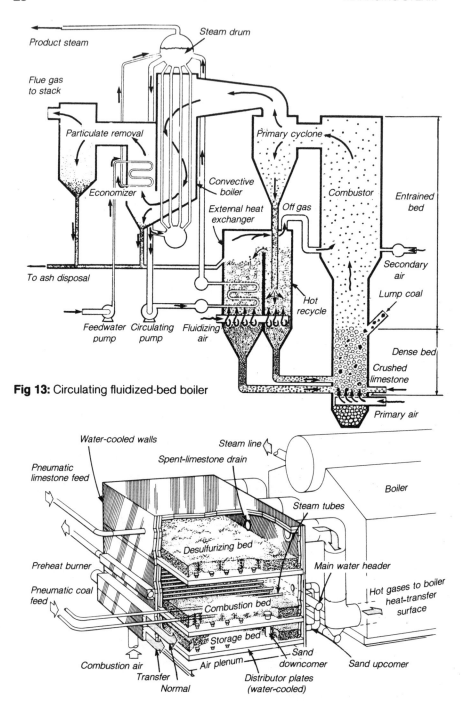

Fig 13: Circulating fluidized-bed boiler

Fig 14: Two-stage fluidized-bed boiler

and gas heat exchange. Ramp-type control is accomplished by regulating the flow of solids through the circulation loop. Rapid step changes, such as a 2:1 turndown rate, can be accomplished by shutting off the air to the solids riser.

The two-stage fluidized bed (Figure 14) is yet another design improvement on the conventional bubbling bed. Here, one stage is for burning the fuel, the upper stage for desulfurizing the flue gas with limestone, the idea being to separate out the two functions so each can be optimized. Steam-generating load is controlled by moving material to and from a storage zone below the combustion stage. One two-stage unit supplier has confirmed an above-95% combustion efficiency and a turndown capability of 30:1.

Commercial fluidized-bed-combustion technology is young and there will undoubtedly be further improvements as the first commercial units continue to gain operating experience. But for generating steam at industrial sites, they are expected to be a significant factor in the marketplace in the years to come.

Hot-water generators

Not all industrial processes require steam, but they still need the heat. For many of them, a hot-water boiler is sufficient.

Like their steam-generating counterparts, hot-water systems are available as packaged, or field-erected, firetube and watertube designs. The first ones were actually modified steam boilers. In fact, the combustion equipment is nearly identical. They are usually classified as low-, medium-, or high-temperature systems, and are rated in million Btu/hr. rather than lbs of steam/hr.

Three important considerations in the design of high-temperature water systems not found in steam-system design are: maintaining the circulating water pressure above the saturation pressure for the water temperature required, providing for expansion of heating-system and generator water, and assuring uniform flow rates under all operating conditions.

Three common pressurization concepts are steam cushion, gas cushion, and mechanical cushion. Steam cushioning is the choice for large-capacity continuously operated hot-water systems—such as those found at airports—because it maintains the tightest control of water temperature. Hot water from the boiler flows into a separate expansion tank where the fluid flashes to steam, maintaining system pressure. Firing rate is controlled by drum pressure. The drum must be located sufficiently above the boiler to permit free vapor release and assure adequate suction head for the circulating pump. Large slugs of cold water returned to the drum will cause the steam cushion to collapse—leading to flashing and water hammer in the piping system.

In the inert-gas pressurization system, a nitrogen blanket is maintained above the saturated fluid at a higher pressure in the expansion vessel (Figure

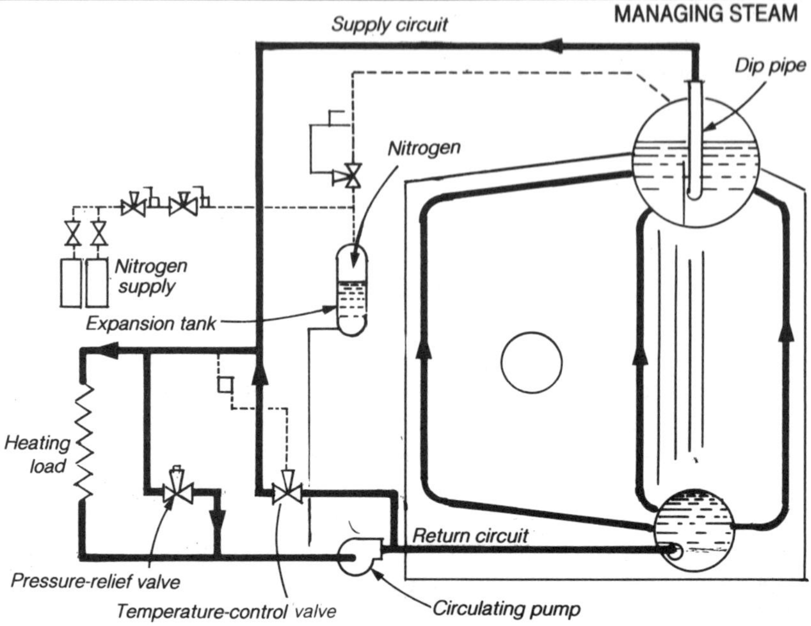

Fig 15: Hot-water generator with inert-gas pressurization

15). An advantage of this arrangement is flexibility in locating the expansion tank. It potentially reduces piping and building costs by avoiding the headroom requirements of the steam-cushion method. Gas pressurization is specified for small systems, and for systems in which wide fluctuations in water temperature are required by the process. Water temperature at the generator outlet is used as an index for combustion control.

Mechanical pressurization, not as popular in industry, makes use of an expansion tank and a pump. As load decreases, the water returning to the generator heats up and expands, increasing system pressure. Once pressure reaches a set point, a dump valve in the expansion tank opens, draining some water to a storage tank. When the firing system responds to adjust the temperature, the pressure decreases and the pump forces water back into the system.

Solar-heating systems

Fired and unfired boilers are not the only ways to generate steam and hot water at industrial and commercial facilities. Since the energy catastrophes of the 1970s, the idea of using solar energy has received added attention. Most of the industrial systems installed today—one survey in late 1981 showed that 42 industrial plants had installed hot-water or steam solar-heating systems—are marginally economic. Most are either sponsored and partially funded by the Department of Energy (DOE), were intended as showcases and not expected to compete with other generating techniques, or were merely token gestures to the great alternate-energy schemes that government officials envisioned for the US at the time. Solar-heating systems for commercial facilities have demonstrated better economics.

Despite high capital costs, the "fuel" is free and another fuel crisis could bring industrial solar systems back into vogue.

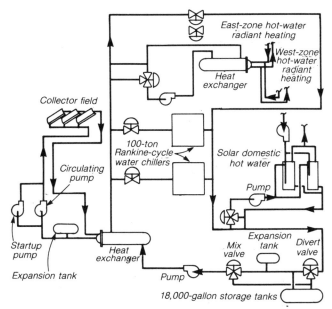

Fig 16: Solar hot-water heating system

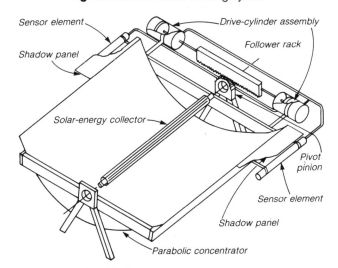

Fig 17: Solar-energy collector plate

Solar-energy thermal systems consist of collectors, conversion equipment, transport facilities, and storage devices (Figure 16). Collectors bring in the sun's energy and concentrate it on the material to be heated, or on a material that is capable of storing the heat. Collectors are further classified as to their composition and geometry—flat plate and tubular mirrors, parabolic mirrors, and parabolic lenses, for example—and/or their ability to track the sun as it moves through its daily cycle (Figure 17). Most of the industrial systems operating today use flat-plate collectors, but they are highly inefficient compared to collectors available now. In the case of a hot-water system, a well-insulated tank is often provided as a storage device.

Steam can be produced at temperatures as high as 700F in the systems being used today, but the average is around 400–500F in most cases. Likewise, hot water over 200F can be produced, but normal temperatures are between 100–200F. Such steam and hot-water conditions are suitable for many of the industrial applications noted earlier.

Chapter three:
Steam distribution in industrial facilities

Once steam is generated in an industrial facility—usually originating at the fuel-fired or waste-heat-fired boiler but also available from steam-turbine exhaust and extraction lines, flashing condensate, or purchased from an external source—it must be delivered at the appropriate temperature and pressure to each respective steam-consuming process. There are essentially two types of consumers: thermal and mechanical. Thermal consumers exploit the heat content of the steam and temperature is the more important parameter; mechanical consumers exploit the thermodynamic work available as steam expands from a high pressure to a lower one. Temperature—especially superheat temperature—and pressure are important. The expansion normally takes place in a steam turbine that drives large pieces of rotating equipment such as compressors, pumps, and fans in large industrial complexes. In this application they are called mechanical-drive steam turbines. Steam also powers turbines that drive electric generators from ones found in the smallest institutional facility to the largest industrial complexes. It is important to note that a highly efficient steam system must be optimized at the source of the steam as well as at the consumer. What follows explains how a steam distribution system is conceived, designed, and operated for maximum efficiency.

Types of users

For thermal utilization, the heat content of steam is transferred to another material either indirectly or directly. In the first use, as implied, the steam contacts the other material. For example, water is sometimes heated by sparging steam into it through nozzles. Other examples are steam humidification, wood pulping where steam is used to break down lignin compounds, steam stripping in refineries, steam cleaning or blasting of metal parts. Direct steam use is wasteful, though, because the condensate is lost or highly contaminated and must be replaced at the source of the steam—resulting in higher feedwater-treatment costs. However, some operations require this use, and the capital cost of direct-contacting heating systems is lower because no heat exchangers, pumps, condensate-return piping, or valving are required.

Indirect use of steam involves transporting it through coils, tube banks, or jackets so that its heat is transferred to a second fluid on the other side of the separating medium. Evaporators, heaters, exchangers, reaction vessels with a heating jacket, dryer drums, and absorption-refrigeration systems are all examples of indirect steam consumers. Steam used indirectly may be prone to contamination, if the steam pressure is lower than the second fluid pressure, but in most cases it is not as severe as in a direct-contact system and the steam can usually be treated for re-use if desired.

Mechanical users convert the work of expansion into mechanical energy which can then be converted to electric power. If a backpressure turbine is used, the exhaust is used for additional thermal requirements.

In large complexes, users are also classified as continuous, intermittent, emergency, and powerhouse. To illustrate: In a refinery, continuous users include strippers, steam reboilers, and turbine drives; intermittent users are tank heaters, winterizing steam tracing, and purging systems during turnarounds; emergency users might include smokeless flares and fire fighting.

Distributing steam

Steam is first generated at a source at a given pressure, high enough to overcome line losses and still meet the delivery requirement of the highest-pressure user, then cascaded through the industrial site (Figure 18). Design of a steam distribution system has two primary goals: 1) delivering steam to all thermal and mechanical consumers at the required temperatures and pressures, flows, and quality, and 2) balancing all requirements so that as little of the steam's energy as possible is wasted.

Once the various steam consumers have been identified, the required flows, temperatures, and pressures are determined by performing energy balances around individual process units, then around the entire plant. Process conditions such as the operating temperature and pressure of the material to be heated, physical properties of the material, especially its heat capacity, and the amount of material must be known. Well-known principles of heat transfer are then applied at the conceptual stage of plant design, but usually there are so many variables constantly changing that actual quantification is difficult. Conceptual plant energy balances should, therefore, be

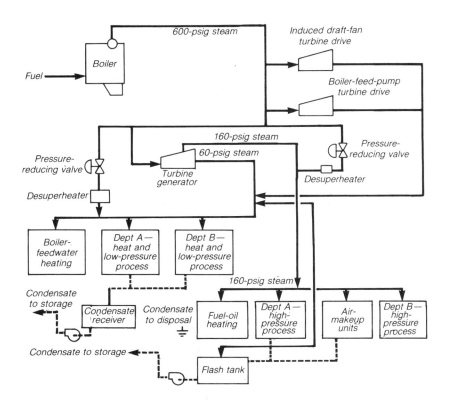

Fig 18: General steam-distribution system

tempered with operating experience and engineering judgement.

An important decision in steam distribution design is at what pressure to distribute the steam. For many building heating systems, this is an easy decision: steam is produced at a pressure sufficient to overcome line losses, and that's all. For facilities that will use steam for a heating system and for process/mechanical, the decision is more complex. In days gone by of cheap fuel, steam was usually generated at a high pressure, then rather indiscriminately reduced in pressure through pressure-reducing valves (PRVs) at the point of use.

It is still advisable to generate steam at higher-than-required pressures to accommodate future plant expansions or key process changes, and because piping costs increase—at least up to the 600 psig/700F steam conditions. Here's why: As steam pressure decreases, its specific volume increases, meaning larger pipe diameters for a constant steam flow to control velocity and line pressure losses. Above 600 psig, heavier-grade pipe is required— offsetting the economics.

Another factor that influences supply pressure is whether to cogenerate steam and electric power. If it proves economical, steam is generated at much higher pressures and temperatures than required by the process and is directed through a steam-turbine/generator, whose exhaust or extraction steam goes into the distribution header. Essentially the turbine acts as a large PRV station with a useful byproduct. Branch PRVs are then used to control pressure to each user. The PRV for the turbine bypass must be sized to handle all necessary plant steam if turbines are shut down.

Deciding whether to cogenerate electric or mechanical power involves a

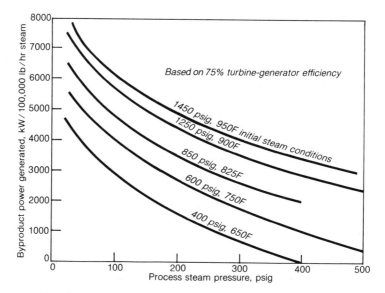

Fig 19: Power available from backpressure steam turbines

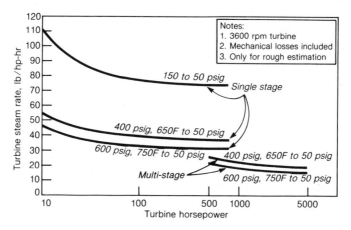

Fig 20: Power vs steam rate for single and multistage steam turbines

lot of economic tradeoffs—more complicated today by electric-power pricing structures from the utility, and the opportunity to sell electric power back to the utility. Many steam turbines are available for these applications. Single-stage units are the least expensive but have lower power-generation efficiencies than multistage units. Large turbines can be supplied with automatic extraction facilities and can operate as extraction or backpressure units (Figure 19). The amount of power available is proportional to the throttle temperature and pressure drop across the turbine (Figure 20), which in turn has important implications for the selection of the steam-distribution-system header pressures. Condensing turbines are usually kept to a minimum when balancing mechanical and thermal-steam requirements. As thermal steam requirements vary, mechanical requirements are met by varying the steam flow to the condenser.

In most large industrial complexes, there will be multiple steam pressure headers such as low pressure (between 10–50 psig), medium pressure (between 50–250 psig), and high pressure (between 250–600 psig). Steam at higher pressures is almost always for electric-power generation.

Multiple header pressures and use of steam turbines make the overall steam balance complex. Often, a computer simulation will aid in optimizing the distribution system, and analyze the effect of upset conditions. The system must be flexible enough to accept load changes without a loss of control, respond promptly to emergency conditions, and handle off-load conditions such as seasonal variations, unit outages, startups, shutdowns, possible electrical-power outages (especially if a mechanical-drive steam turbine is used as a backup to an electric motor).

Two key parameters must be controlled: excess steam flow through pressure-reducing stations, and excess steam vented to atmosphere. Both waste the mechanical and thermal energy inherent in the steam. Simulations reduce the risk of under- or oversizing critical components such as PRV stations, desuperheating stations, and vent silencers. Another benefit is balancing the use of steam turbines and electric-motor mechanical drives. For example, when exhaust steam from turbine drives exceeds process requirements, the excess has to be vented. Depending on electric-power costs and process-reliability considerations, motors may be the wiser choice.

Once the individual consumers have been identified, a site plan showing the location of the consumers and each one's demand should be drawn up, as well as all PRVs and turbines. Piping routes can then be specified. Steam lines are generally laid overhead in industrial facilities, because underground lines, where practical, are limited to low-pressure steam. Steam-line sizing is a standard procedure and the nomographs and tabulated data are available in the appendices.

All steam lines should be insulated for personnel safety and to reduce heat losses. Calcium silicate insulation with aluminum jacketing is the one selected most often.

Auxiliaries

Piping auxiliaries, especially valves, should be readily accessible for periodic maintenance. For long steam lines, pockets should be avoided or suitable drains provided via steam traps and drains. Sufficient valving should be included to isolate sections of pipe with minimum impact on the process. Some refineries recommend a valve and blank in each steam line entering or leaving a process unit, and for off-site steam leads. Large valves should be completely opened and closed at least once a year to prevent possible freezing in position. Repacking programs will minimize any loss of steam due to leaks in the packing. Other maintenance procedures are described in later chapters.

Steam's image is changing

In today's energy scenario, the value of steam is being determined more rigorously to better evaluate waste-heat-recovery projects. One method of assigning costs to steam makes use of a fuel-value parameter. Absolute fuel value (AFV) is the amount of fuel (in Btu) needed to generate a given quantity of steam. Relative fuel value (RFV) is the ratio of the AFV of low-pressure or medium-pressure steam to a reference—the AFV of high-pressure steam. These can be easily converted to dollar values as well. Such a method accounts for parameters such as turbine efficiency and boiler efficiency and allows comparisons among steam extracted from condensing turbines, exhaust steam from a backpressure turbine, and steam from reducing stations and condensate flash drums.

A corollary to today's emphasis on valuing steam is the importance of knowing how much steam is being used at each consumer. This can be a cornerstone to establishing accountability for steam consumption by specific plant areas. It also aids in determining the energy cost per quantity of product. Steam metering and accounting also helps in tracking down major steam leaks. Good practice is to conduct the steam survey once a month, with results reported to individual departments. Accounting procedures are especially amenable to today's computer hardware and software. Some plants have installed dedicated computer systems for metering, calculating, record-keeping, and reporting.

Hardware and instrumentation for steam metering are discussed in later chapters.

Condensate-return systems

The third major subsystem of a complete steam-distribution system is the condensate-return facilities. Each pound of condensate not returned to the steam generator is a pound of raw water that requires expensive chemical treatment. In addition, condensate represents sensible heat that, if recovered, improves the overall energy efficiency of the steam system. Inefficient older facilities tend to discharge much of the heat in condensate as flash steam. New state-of-the-art systems return as much condensate as possible at the highest possible temperatures.

Two limitations on achieving maximum recovery are the quality of the condensate and the capital costs involved. Not all condensate is reusable. Even steam used indirectly can become contaminated. For example, condensate with oil in it could severely damage the steam generator if the condensate/oil got mixed in with boiler feedwater. A large facility might have separate condensate-handling facilities for contaminated and non-contaminated streams.

On the other hand, some condensate streams are too small to recover, especially in view of the capital costs and associated maintenance costs involved for pumps, tanks, piping, valving, and steam traps.

The two common types of condensate-return systems are pressurized (Figure 21) and non-pressurized. In the former, steam is used as the driving force to move condensate into a storage tank. There will be a backpressure on steam traps, process-heat exchangers, and steam-consuming equipment that is equal to the friction loss plus the receiver pressure. This equipment must be sized to handle the backpressure, and the inlet steam pressure must be correspondingly higher. Except for the condensate traps, a pressurized sys-

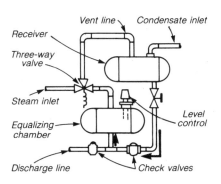

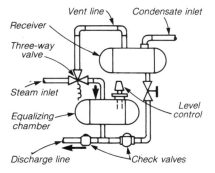

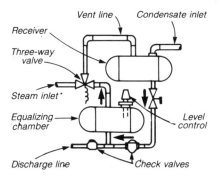

Fig 21: Condensate-return system. Fill cycle, top left; equalization cycle, top right; vent cycle, bottom

tem requires minimal maintenance because pumps and controls are not needed. But most condensate-return lines must be sized to handle two-phase flow (Figure 22).

Pressurized systems can be further classified as open, closed, or flash (Figure 23). In an open system, the condensate-storage tank is vented to the atmosphere. It can handle hotter condensate than a more conventional pump system, and handles condensate from multiple sources at different pressures.

A closed system is not vented to atmosphere; its primary advantages are that it eliminates unrestricted flashing, and can handle condensate at any

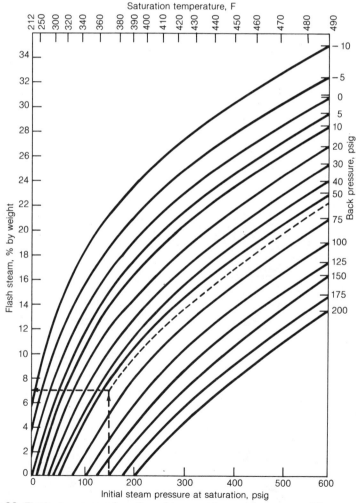

Fig 22: Flash steam as a function of temperature and pressure difference

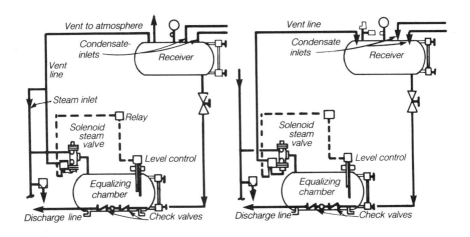

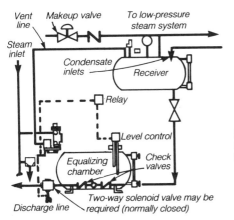

Fig 23: Pressurized condensate-return systems. Open type, top left; closed type, top right; flash type, bottom

temperature. Its disadvantage is that it can handle condensate from only one pressure source. Otherwise, the receiver stabilizes at some positive pressure and can cause flooding of the lower-pressure return line.

A flash system can be considered a combination open/closed system. It has a low-pressure takeoff on the receiver. Mixed pressures from multiple returns can be handled, and condensate can be returned to the boiler room at whatever temperature it reaches the equalizing chamber. Also, heat can be recovered that would otherwise be lost through indiscriminate flashing.

An unpressurized system, the more conventional of the two, has traps that discharge into a centralized receiver tank, then the condensate is pumped back to the powerhouse. It is common practice to vent the condensate receivers to atmosphere, resulting in some loss of flash steam.

Chapter four:
Cogeneration

Cogeneration—the simultaneous production of steam and mechanical or electric power—is a growing trend in the industrial energy market. Yet there is nothing really innovative technologically about cogeneration. Industrialists used to cogenerate more than 50% of their energy needs in the early part of this century, until purchasing electric power from the local utility became cheaper and more reliable. The equipment and systems discussed earlier all apply to the cogeneration field.

Instead, revived cogeneration activity reflects the breakdown of institutional barriers that have inhibited efficient energy production, and the changing state of energy politics and economics. Purchased electric power is no longer cheap. Utilities face ever-increasing problems with building the massive fossil-fired and nuclear generating central stations that, in the past, took advantage of the economies of scale to help provide cheap power for sale to industry. Premium fuel price escalations of the 1970s dealt cheap utility-supplied power another blow.

Industrial steam-system engineers who generate their own electric power insulate themselves from the cost overruns and other problems that make secure, steadily priced utility power an uncertainty for the operating life of the plant. Perhaps the strongest prop holding the cogeneration field up is government legislation. Utilities are now obligated to negotiate in good faith a contract to buy back excess electric power generated at an industrial or commercial site. One of the past hindrances for industry to cogenerate was the mismatch between its steam and electric-power loads. Now, if the utility buys excess power, then industry can generate the maximum steam demand of its plant and produce electric power as a saleable byproduct. In some cases, the utility will have to pay the full avoided cost of what it would cost to generate that same amount of power itself.

Not all plants have a steam/electric-power demand that results in excess electric power for the thermal load demand. Such a plant is still a cogenerator. By generating electric power it reduces its costs by lowering purchased power costs, rather than from increased revenue from the sale of excess power.

The concept of cogeneration is now being applied with systems as small as 100 kW and as large as 500 MW. First order of business for industrial plants considering this approach is to decide whether steam and electric-power loads are amenable to a cogenerate scheme.

To evaluate the feasibility of a cogeneration project, compare the potential savings in operating costs with any capital costs that might be required. Keep in mind that operating costs have an energy component and equipment operation and maintenance component, the former being dominant when looking at profitability. Calculations of energy cost involve taking into consideration plant-performance characteristics relative to anticipated plant load; the better the estimate of thermal load, the more accurate the estimate of energy cost will be.

Normally it is not economical to design a cogeneration plant to produce

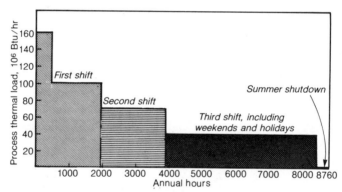

Fig 24: Process thermal-load curve, top; Heating-plant thermal-load curve, bottom

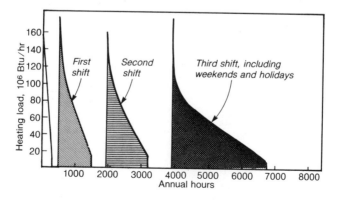

all of the required thermal and electrical energy. Peak energy supply is better obtained from sources such as supplementary firing of on-site boilers and electric-power purchases from the utility. But if the right agreement can be reached with the utility, then it may pay to produce electric power beyond the needs of the plant and sell the excess.

Most plants do not have constant thermal or electrical loads. Both tend to vary with the time of day (depending on the shift, for example), the day of the week (weekday or weekend day), or the season (heating or non-heating, high production or low production).

Because the cost of energy production varies between sources, the total amount of energy from each source should be included in the cogeneration analysis. For that matter, the cost of energy from a single source varies during peak or non-peak hours, or at full- or part-load operation. For example, the efficiency of a combined-cycle plant may decrease by 4–5% at 75% load and by up to 15% when operating at 50% load.

Annual energy-consumption rates and totals, as well as simple averages, produce misleading or incorrect results when they are used to determine the costs associated with an energy source in the cogeneration analysis. Consider a conventional thermal-load curve (Figure 24). Instead of assuming the area

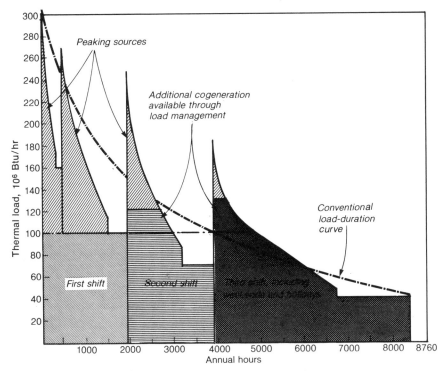

Fig 24 (cont): Detailed thermal-load curve

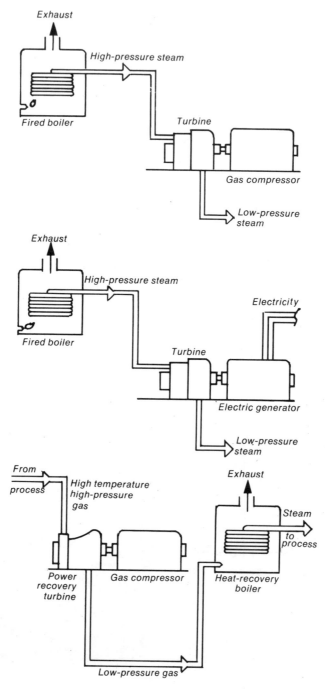

Fig 25: Topping cogeneration cycles, top and center; bottoming cycle, bottom

under the curve is equivalent to the total annual energy consumed or produced, divide the year into periods in which simple thermal-load patterns can be established. Load for each period may either be constant or vary as a function of an independent parameter—such as ambient temperature or process needs. After thermal-load increments have been established, a more accurate annual load-duration curve can be developed and used in the analysis.

Once the cogeneration option has been established as an economical alternative, the designer must choose among a seemingly endless variety of topping and bottoming cycles (Figure 25): gas turbines with heat-recovery steam generator, combined-cycle plant with steam extracted for process use, diesel-engine generator with heat-recovery steam generator, fired boiler with extraction/condensing, backpressure steam turbines (Figure 26), and so on.

Though industry has preferred the use of backpressure steam turbines, a condensing turbine with controlled extraction may be the economical choice in some cases. Condensing turbines feature a higher electric-power output for a given steam flow than a backpressure unit. The high-pressure section of the turbine performs the same as a backpressure unit, and the extraction valve maintains the process-steam pressure.

Choice between the two will ultimately depend on the negotiated price of power sold to the utility and the thermal/electric load balance at the plant. For example, backpressure turbines show peak electrical production during the winter months because they follow the steam load directly and more steam is usually needed in the winter. Conversely, a single controlled-extraction condensing machine produces the most power during the summer months because steam demand is lowest and much more steam flows on to the condensing section of the turbine, instead of being consumed by the process. Utilities—because of the usual summer capacity problems—may be more amenable to a higher negotiated price for the additional power in the summer.

Gas-turbine/heat-recovery steam generators make flexible cogenerating systems mainly because of the relatively large amount of heat energy available in the exhaust compared to the power output. Efficiency of the gas turbine is mostly dependent on the pressure ratio and the gas-turbine-inlet temperature. Turbines with low pressure ratios are less efficient in generating electric power, but have more recoverable heat in the exhaust.

Two options increase the flexibility even further. Supplemental firing of the steam generator helps boost the thermal-heat (steam) to electric-power ratio, but tends to decrease the overall efficiency because of the fuel consumption. Nevertheless, it is frequently economical because the thermal energy provided uses less fuel than a conventional separate ambient-air fuel-fired unit. An extraction/condensing steam turbine permits generating additional power if the process-steam demand is less than the generating capacity of the turbine.

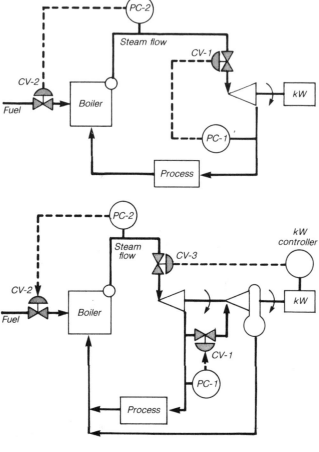

Fig 26: Cogeneration equipment options. Boiler/backpressure turbine, top; boiler/extraction-condensing turbine, center and bottom

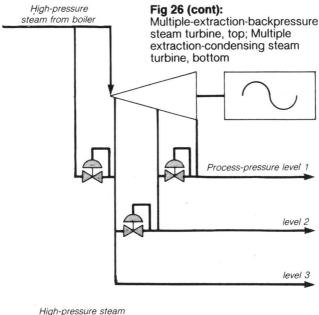

Fig 26 (cont):
Multiple-extraction-backpressure steam turbine, top; Multiple extraction-condensing steam turbine, bottom

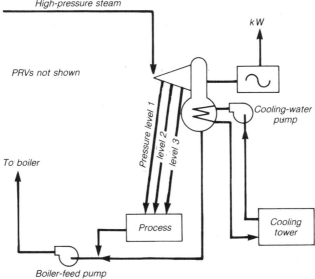

Another way to achieve flexibility in cogeneration is to inject superheated steam into the gas turbine to augment the hot expansion gases and take advantage of higher overall mass loadings without changing the turbine-inlet temperature. This arrangement is especially suited to industrial sites where the steam and electric-power loads are out of step with each other. When steam is not required by the process, the injection feature maintains turbine output without having to decrease it to follow the decreasing steam demand—or vent the steam not required at that time. When steam is not injected into the gas turbine, the turbine still operates at its standard performance level.

Diesel-engine/heat-recovery steam-generator cogeneration systems are used extensively in the under-10-MW size range, and are especially popular with institutional and commercial users, though gas-turbine-based systems are making inroads with them as well. In many cases, packaged systems are available. Quantity of heat recoverable from a engine is less than that from a gas turbine, because the exhaust flow is lower.

In most reciprocating combustion engines, heat is reclaimed from the lubricating system, jacket cooling-water system, and engine exhaust as either low-pressure steam or hot water. Of these, lubricating-oil heat recovery is the least significant. Steam or hot-water production is generally controlled by passing a portion of the exhaust gases around the boiler tube bank and out through a gas-bypass-valve assembly.

Engine-jacket water-cooling passages such as those in the block, heads, and exhaust manifolds remove about 30% of the heat input to the engine. A pump-driven water-circulation system is usually used and the hot water is then available for process loads, space heating, absorption-refrigeration chillers, domestic hot water, and so on. To keep reliability high at multiple engine installations, often a heat exchanger is provided for each engine and the heat in the engine cooling water is transferred to a secondary heat utilization system. Also, to protect the engine from sudden cooling, a hot-water-storage tank is sometimes provided that serves as a buffer, providing heat at a very high rate for short periods of time. Another way of protecting the heat-recovery steam system from instantaneous load changes is by flashing the recovered hot water into steam. Adequate static head above the pump inlet is essential to prevent a change in pressure and steam from forming in the engine.

Some engines use natural convection, ebullient cooling where water is circulated by gravity head into the bottom of the engine; heat from the engine causes steam bubbles to form, lowering the fluid density and causing it to rise and discharge into a flash tank, where the steam is released and the water recirculated to the engine.

Other conservation projects

It is important to note that cogeneration at any industrial or commercial site will compete heavily with conservation projects that lower the required steam demand even further. Industry has made great strides recently in reducing the amount of steam it uses at existing plants. In addition, new processes that are far more energy efficient than their predecessors are expected to come on-line should demand for industrial and manufacturing goods warrant wholesale expansion of facilities.

. Two examples of conservation projects that go beyond housekeeping measures are recovering heat from boiler blowdown and upgrading low-pressure waste steam for further process use.

Blowdown of boiler-drum water is necessary to maintain the solids concentration in the drum to within specified limits. Rate of blowdown depends on boiler operating pressure, quantity and quality of makeup water, type of pretreatment equipment, steam-purity requirements, and boiler design. Some industrial boilers require continuous blowdown as high as 10%. In many cases, heat recovery from continuous boiler blowdown is economical. Further, local environmental laws may require that discharges to municipal sewers not exceed 100F, so some cooling would be required in any event.

Saturation pressure in a boiler determines the amount of heat that is wasted in the blowdown. For example, about 240 Btu/lb of blowdown is recoverable from a 100-psig boiler, and about 600 Btu/lb from a 1900-psig unit, both based on cooling the blowdown to 100F. Percentage actually recovered is based on the type of system installed, but in most cases will fall between 65–95%.

A typical heat-recovery system consists of a flash tank, a heat exchanger, and a float control valve. Continuous blowdown from the boiler is accomplished by the flow-control valve, which allows bleeding of drum water at a set rate. For multiple-boiler installations, blowdown lines can be linked to a common header before entering the flash tank. Blowdown valves should be placed as close to the flash tank as possible.

The blowdown drain from the tank goes through a heat exchanger to recover the sensible heat of the water. Typically, the heat exchanger is sized to reduce the temperature of the blowdown drain to 20 deg F above the inlet temperature of the fluid being heated.

Virtually all industrial plants can locate sources of low-quality steam being exhausted to atmosphere, wasting the latent heat of vaporization. One

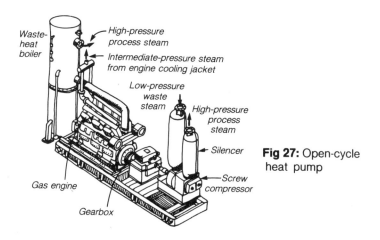

Fig 27: Open-cycle heat pump

way to take advantage of the heat is to recompress the steam in an open-cycle heat pump (Figure 27). In some cases, it may be more economical to do this than generate the steam from a direct-fired boiler. Most steam-recompression methods require that the steam be cleaned of any contaminants prior to being recompressed in a centrifugal and/or axial compressor. Using a screw compressor might avoid this problem. Economic performance of such a system usually depends primarily on the prime mover that drives the compressor—electric-power cost for a motor-driven unit or fuel costs for a gas-turbine-driven or steam-turbine-driven unit.

SECTION II:
UTILITY STEAM SYSTEMS

Chapter Five:
Generating steam
for electric power

In contrast to industrial facilities, electric utilities produce steam for one primary purpose—to generate electric power as efficiently as possible. Barring a few auxiliary uses—such as startup steam for warming or flushing lines, or to power mechanical-drive turbines—the bulk of the steam is directed through large turbine/generators. Because of this overriding objective, utility steam requirements involve higher temperatures and pressures than industrial steam requirements.

Capacities for utility steam generators go as high as 10 million lbs/hr. Also, the quest for the highest fuel-to-electric-power conversion efficiency dictates the need for superheated steam at the highest temperatures and pressures that can be handled. Steam pressures reach 3800 psig and temperatures usually top out at around 1050F. There are indications from several utility R&D houses that the next generation of utility steam systems will go beyond these limits to increase efficiency.

With the large flowrates and sophisticated steam conditions, steam-handling equipment is not only greater in size, but involves very specialized engineering so that reliability, operating life, availability, and efficiency are not compromised. On top of the engineering challenges comes the need for specialized water treatment to prevent deposition and corrosion problems in the steam generator.

Interaction of the steam generator and the major steam consumer—the steam turbine/generator—also separates utility plants from industrial plants. To optimize the overall power cycle, steam is returned from the turbine at intermediate stages and reheated in the steam generator, sometimes twice. Lesser amounts of steam are extracted from the turbine to preheat the feedwater recycled to the steam generator and to power boiler-feed pump turbines and other mechanical drives throughout the plant.

The heat source for producing steam is also a factor when comparing central stations and industrial steam plants. Fuels used are similar to those used in the industrial boilers mentioned earlier, but firing systems may be quite different. More importantly, some utilities make use of nuclear-powered steam generators. This involves a level of engineering not known in industrial powerplants, especially in meeting strict regulations for safety and reliability.

Finally, the whole issue of steam-system control is changing in US powerplants.

In the past, large fossil-fired steam generators were operated base load. Control schemes were relatively straightforward. Now that nuclear units have come on line in the last ten years, they are baseloaded at many utilities, while the baseload-designed fossil-fired units are cycled through daily and weekly starts and stops. This change in duties has complicated the control of utility steam systems.

Changes in the energy economics also impact greatly on the area of control. When fuel was inexpensive, it was easy to design plants with comfortable margins for the sake of added reliability, even if efficiency was sacrificed somewhat. Now, utilities are interested in optimizing station efficiency in part through on-line control and monitoring of the steam/condensate circuit, and by re-evaluating design details of the steam's path through the turbine.

What follows is a review of utility steam systems of today and tomorrow, the steam generators, turbines, and auxiliaries such as feedwater heaters, economizers, and condensers, and a look at some of the important operating and design issues with today's systems.

Fossil-fired steam generators

Steam generators for utility service can be classified by the type of heat source, fossil-fired or nuclear, the steam cycle (whether it's a subcritical or supercritical system), and whether it's a drum-type of unit or a once-through unit.

In the US, the fossil-fired subcritical watertube drum-type boiler (Figure 1) has predominated over the last several years. Steam pressures are either 1900 psig or 2600 psig. The reason for its predominance is the better reliability compared to the supercritical once-through unit, although most boiler vendors and consultants feel that the supercritical unit's reputation suffered from poor performance during its commercialization process in the 1960s. The roots of this have been largely corrected. Except for the increase in size and the need for superheated and reheated steam, the subcritical utility

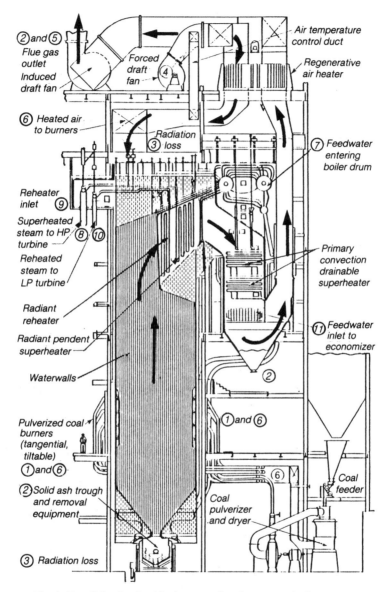

Fig 1: Fossil-fired subcritical watertube drum-type boiler

boiler design is similar to the large industrial boilers mentioned earlier. A saturated steam/water mixture continuously circulates under either forced or natural circulation. Steam that separates in the drum is then superheated. Likewise, steam from the turbine is reheated.

Superheaters and reheaters are heat-transfer surfaces which raise the temperature of steam above its saturation point. Two reasons for superheating are to take advantage of the thermodynamic gain in efficiency, and to ensure that the steam is dry and does not condense in the last stages of the turbine. There is an approximate gain of 3% in overall efficiency for every 100F of superheat (depending on steam pressure as well).

Both heat-transfer surfaces are distinguished by their locations in the boiler and by the predominant mode of heat transfer. In convection superheaters and reheaters, located downstream of the furnace, steam temperatures rise as steam flow increases because mass flow on the gasside increases faster than the steam flow. The opposite happens in radiant superheaters, which are located in the furnace. Here, steam temperature drops as load increases because available heat does not increase at the same rate as steam-

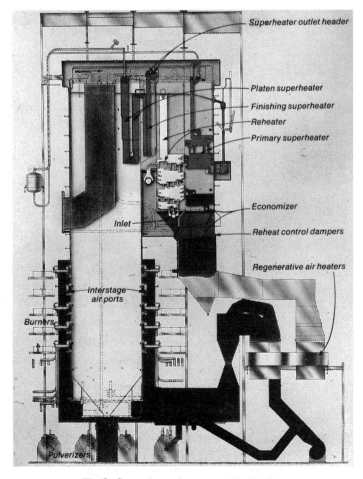

Fig 2: Once-through supercritical boiler

mass flow within the tubes. Therefore, to hold the temperature reasonably constant as steam output varies, superheater or reheater surface is often split between the radiant and convective sections.

Steam temperature can be further controlled from the steamside or the fireside. On the steamside, attemperation or desuperheating is a technique whereby the steam temperature is lowered—generally by spraying pure water into the steam.

Steam-drum internals are an important aspect of drum-type boilers. They are designed to keep water droplets and the contaminants they contain from entering the steam turbine. One drum design makes use of cyclones and scrubbers above the cyclones along the length of the drum that separates water droplets by gravity. Pure water is then sprayed into the scrubbers to wash out the contaminants, such as silica, in the vapor phase and return them to the water. Coupled with the internals, drums for today's units have become larger to provide a greater margin for handling upset conditions.

Once-through supercritical units (Figure 2)—which can be fired with all fossil fuels—operate at much higher pressures and do not have a conventional steam drum. They operate at 3800 psig and 1000F—above the critical pressure of 3206.2 psia where steam and water coexist at the same density. When this mixture is heated above the saturation temperature of 705.4F, dry superheated steam is produced. These conditions offer about a 1.6% gain in overall plant efficiency compared to the 2400-psig drum-type steam generator.

The complete steam-forming process in a once-through can be thought of as being carried out in a single tube. Feedwater enters at one end and superheated steam emerges at the other. In practice, a large number of individual tubes discharge into a common header.

Once-through steam generators pose special problems in steam temperature control and water treatment. In one type of supercritical unit, temperature is maintained at the setpoint by attemperation before entering the final superheater. Steam is directed through the high-pressure portion of the turbine, then is reheated in two sets of reheater surfaces, and sent back through the intermediate- and low-pressure turbine sections.

Since once-through units have no natural recirculation, there must be a provision for supplying a heat-absorbing medium to the furnace walls to prevent overheating at low loads. One design makes use of a mixing vessel and a recirculation pump. Fluid from the mixing vessel is automatically pumped through the heat-absorbing surfaces. This is called a floating recirculation system. Above 50% load, it doesn't contribute to the steam/water flow. Below 50%, when feedwater flow alone is not sufficient to protect the furnace walls, the recirculating pump kicks in.

An alternative method is to use a bypass. Steam flow is diverted around the turbine into a flash tank and the separated water is used to maintain the required mass flow in the boiler tubes.

Nuclear steam systems

The heat source for a nuclear steam system is uranium dioxide rather than fossil fuel. On the steamside, the essential difference between fossil and nuclear steam generation is the lower steam conditions. In the largest boiling-water reactor (Figures 3 & 4), up to 17 million lbs/hr of steam is generated at about 1000 psia and 550F. Key components are the cylindrical-reactor pressure vessel, steam-separating and drying equipment, feedwater distributor, and internal recirculation system.

The nuclear "furnace" or core comprises hundreds of fuel-rod assemblies immersed in the water. Water is heated and becomes a two-phase mixture which enters a plenum directly over the core. Saturated steam is separated

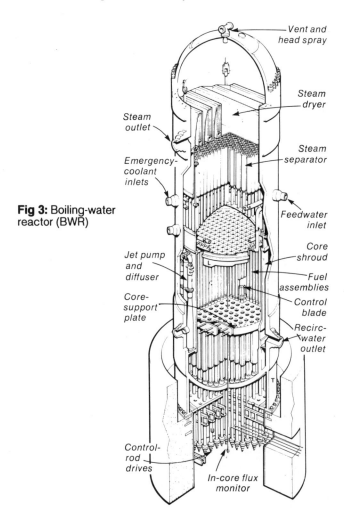

Vent and head spray

Steam dryer

Steam outlet

Steam separator

Emergency-coolant inlets

Feedwater inlet

Fig 3: Boiling-water reactor (BWR)

Core shroud

Jet pump and diffuser

Fuel assemblies

Core-support plate

Control blade

Recirc-water outlet

Control-rod drives

In-core flux monitor

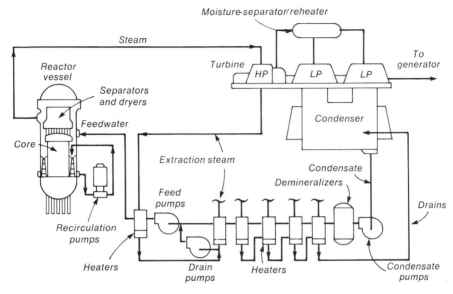

Fig 4: BWR steam system

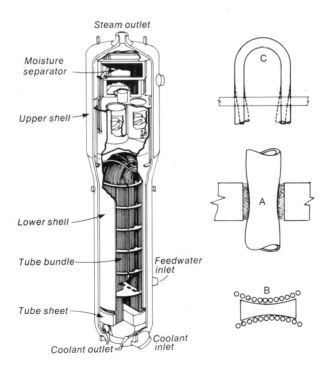

Fig 5: Pressurized-water reactor (PWR)

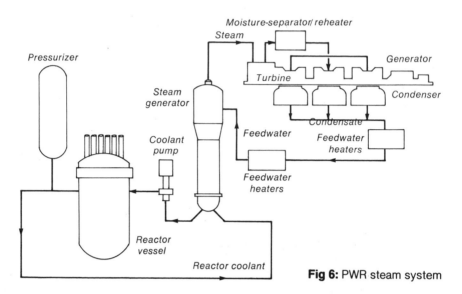

Fig 6: PWR steam system

from the water and passes through a dryer to remove moisture while the water is recirculated back to the reactor vessel. Steam rate is controlled by positioning control rods vertically within the core. Further steam-rate control is provided by varying the recirculation of water within the core.

Another nuclear-power cycle makes use of an indirect steam cycle. It is classified as either a once-through or recirculating pressurized-water reactor (PWR) (Figures 5 & 6). A primary loop provides water at 2250 psia and between 540–600F. This water flows to the steam generator which—in the recirculating case—consists of U-tubes, an evaporator section with inverted U-tubes, and an overhead steam drum and moisture separator. Primary water enters at the bottom, flows through the U-tubes, and exits at the bottom as well. Feedwater, on the other hand, enters above the tube bundle and flows on the shellside. Steam exits at the top at about 550F and 1050 psia.

The once-through steam generator (Figure 7) features a straight-tube design. Pressurized cooling water enters from above, flows down through the tubes, and out the lower end. Feedwater, preheated by aspirated steam on the way down a peripheral annulus, flows up on the shellside of the unit. It is slightly superheated on the way up the shell, reverses direction again, then exits through nozzles located slightly above the feedwater inlet ports.

Auxiliaries

Utility steam systems are equipped with a variety of auxiliary components to optimize the overall steam-to-electric power cycle: economizers, feedwater heaters, condensers, and deaerators.

In fossil-fired systems, economizers are located downstream of the super-heater in the convection section to extract more heat from the flue gas by preheating the feedwater. Integral economizers are supplied as part of the convective surface for most utility boilers. Extended heat-absorbing surfaces are needed because of the low temperatures involved.

The feedwater temperature is further raised by feedwater heaters (Figure 8) that use steam normally extracted from the turbine. Depending on the size

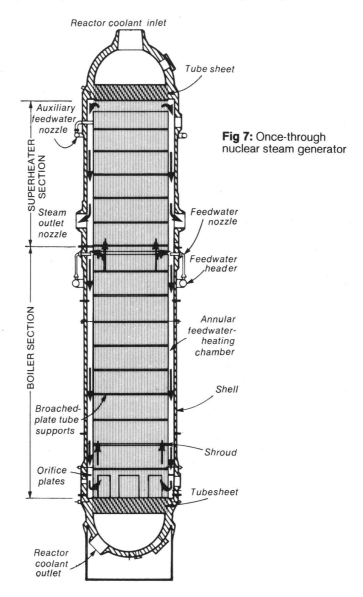

Reactor coolant inlet

Tube sheet

Auxiliary feedwater nozzle

SUPERHEATER SECTION

Fig 7: Once-through nuclear steam generator

Steam outlet nozzle

Feedwater nozzle

Feedwater header

Annular feedwater-heating chamber

BOILER SECTION

Shell

Broached-plate tube supports

Orifice plates

Shroud

Tubesheet

Reactor coolant outlet

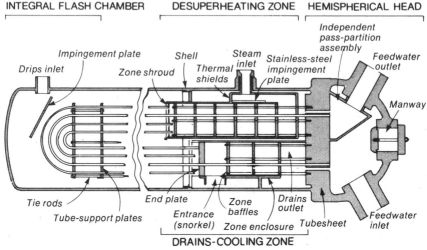

Fig 8: Feedwater heater

of the plant, the duty (peaking, baseload, cycling), and power cycle (nuclear, fossil), up to seven or eight stages of feedwater heating may be used.

Feedwater heaters are classified as low, intermediate, or high pressure depending on the location within the cycle. Low-pressure units are located between the surface condensers and the deaerating heater; intermediate-pressure heaters between the boiler-feed booster pump and the boiler-feed pump; high-pressure units downstream of the boiler-feed pump. Feedwater flows inside the tubes at pressures as high as 5000 psig in supercritical fossil cycles. Steam flowing on the shellside ranges in pressure from vacuum to above 1200 psig for high-pressure units.

Deaerators are also used to increase thermal efficiency and, more importantly, to keep oxygen out of the steam system. Three types are available: spray, tray, and combination spray/tray units (Figure 9). Spray deaerators are steam-filled enclosures into which the feedwater is sprayed. In the process, it is heated and scrubbed of impurities. Entering fresh steam agitates the water to liberate any remaining impurities. Tray deaerators direct feedwater into a series of cascading trays. As it cascades by overflow or through small holes, steam surrounding the trays heats and deaerates the water. Tray/spray units combine these principles.

Condensers (Figure 10) serve two purposes. They lower the backpressure on the steam turbine to maximize electric-power generating efficiency, and they recover the valuable condensate for reuse as feedwater. It is operated vacuum tight with cooling water on the tubeside and the turbine-exhaust stream and other minor streams on the shellside. Arrangements include one- and two-pass units with from one to three shells that operate at equal or different pressures depending on the turbine size, expected thermodynamic performance, and capital cost.

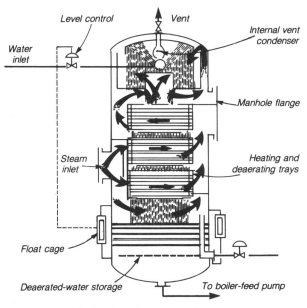

Fig 9: Spray/tray-type deaerator

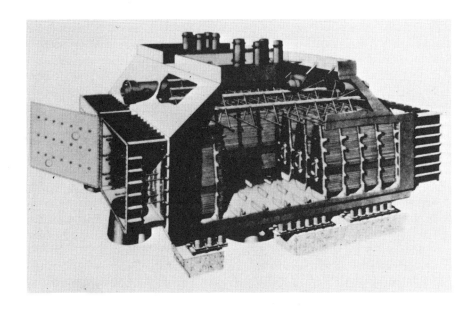

Fig 10: Large powerplant condenser

Steam turbines

Although steam turbines come in an almost infinite range of sizes, the ones used at most US utilities driving generators are large condensing and reheat units (Figure 11). They are often rated well above 1 million horsepower. Condensing steam turbines exhaust steam at less than atmospheric pressure. The vacuum is maintained with a condenser and air ejectors of either the steam-jet or mechanical-pump type. Shaft-sealing steam is applied to the turbine glands to prevent air from entering the seals and destroying the vacuum. Condensing turbines are more efficient for generating electric power than the backpressure ones commonly found in industry.

Reheat is another feature which boosts electric-power-generating efficiency. Main steam from the boiler enters the turbine and exhausts from one or two intermediate stages, recirculates back to the boiler where its temperature is raised, then returns to the next lower turbine stage for further expansion. If there are two intermediate stages, then the cycle is called double reheat.

Large turbine/generators can have up to a dozen bleed points for feedwater heating and other duties. The pressure of the extracted steam varies with the turbine-shaft load—called non-automatic extraction. Feedwater-heating service does not generally demand careful control of the bleed-steam pressure.

Most turbine/generators receive steam at 2400 psig or 3500 psig in fossil-fired plants, but closer to 1000 psig in nuclear plants. Temperatures differ as well. Nearly all newer fossil units are designed for a nominal 100F steam temperature at the throttle and reheat lines. Some units designed exclusively for baseload service may have temperatures up to 1050F at the throttle, while those expected to see much cycling duty could see 1000F at the throttle and reheat inlet. On the other hand, nuclear steam turbines are more likely to see temperatures of 550–600F.

Utilities also use steam turbines as drives for pumps and fans in both fossil and nuclear plants. Especially in coal-fired plants above 400MW, steam turbine drives meet variable pump loads reliably and improve the overall cycle efficiency. Electric-motor drives tend to be more economical for plants whose boiler-feed-pump power requirements are less than 8000 hp.

Steam for boiler-feed pump drives is generally obtained from the main unit crossover during normal operation and may either exhaust to the main condenser or to a separate condenser. Flow through the pump drive turbine parallels steam flow in the main unit. Because the main-turbine exhaust losses are reduced, there is a net gain in thermodynamic efficiency. Sometimes, when the unit runs at low loads, a backup source of steam is used to drive the boiler-feed pump.

Steam drives the induced-draft (ID)-fan turbines in large plants as well. The power requirement for fans decreases more rapidly than for boiler-feed

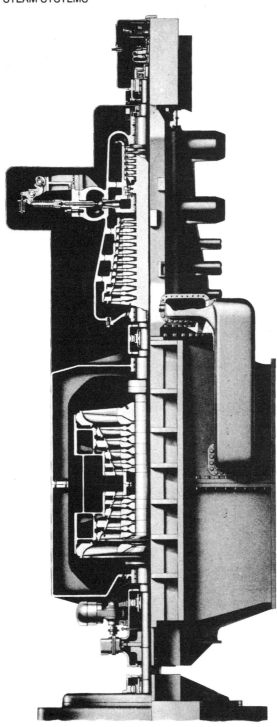

Fig 11: Utility steam turbine/generator

pumps, so the steam-turbine drive here has a broader operational range. In fact, the steam exhaust from the fan drive can sometimes be used to accomplish some air preheating in an exchanger ahead of the air-to-air preheater. In nuclear plants, steam turbines are also used to drive feedwater pumps and emergency core cooling pumps. Many of the units are single-stage solid-wheel machines that range beyond 1000 hp.

Important issues today

During the 1950s, utilities engineers were predicting that steam temperatures and pressures would advance to the 5000–7000-psig range and the 1100–1200F range to further improve overall station efficiency.

This trend never materialized. Utility steam conditions have remained where they are today for the last 20 years. There is some revived interest in utility R&D organizations to advance steam conditions and develop the equipment to handle it, but no commercial endeavours are expected for the next several years at least. In the meantime, improvement focuses on the combustion and the steam/condensate loops.

On the combustion side, the interest is in using coal and reducing the use of premium fuels such as natural gas and oil, but in an environmentally safe manner. To this end, the fluidized-bed steam generator is being commercialized for utility service. Already, there are several small utility units under 100 MW operating. Load following is one of the major issues with fluidized-bed technology, as is the need for steam-generating tubes immersed in the fluidized solids, at least in bubbling-bed designs. The conventional combined-cycle powerplant may become an important way to use coal and coal-derived liquids and gases, or combined with an atmospheric-fluidized-bed air heater that burns solid coal and heats air that is subsequently expanded in a gas turbine. As its name implies, combined-cycle means that a hot-air cycle drives a gas turbine, and steam—usually produced from the gas-turbine exhaust heat—drives a steam turbine/generator. Already at several US utilities, combined cycles are firing premium fuels with a 10–15% improved fuel-to-electric power efficiency over a conventional fossil-fired plant with SO2 scrubbers. One drawback is that the control scheme becomes more complicated because two interdependent power-generating cycles have to be regulated simultaneously.

The first integrated coal-gasification combined-cycle powerplant in the US recently began operating in late 1984, and undoubtedly the experience will indicate what the future of this scheme will be for US utilities. There is also interest today in converting the many gas turbines at utilities installed for peaking duties into combined-cycle powerplants. This would be a way to add baseload power capability in small increments when load growth picks up again. In addition, several US utilities are looking into a pressurized-fluidized-bed combustor/combined-cycle system. Here, the combustion air from the furnace, after being cleaned of ash and other material, is under pressure and is available to drive a gas turbine. Steam, which is produced in tubes placed in the bed, drives a turbine/generator. There is a marked gain in power-generating efficiency with this scheme over the atmospheric-fluidized-bed arrangement.

One of the most important issues to utilities today is converting fossil-fired units—designed to operate baseload—into cycling units. Under baseload operation, all control valves are wide open most of the time so there is not throttling loss across the valves. When cycled, steam to the turbine is controlled with the main control valves on the turbine steam chest outlet, while the boiler is maintained at full pressure. Throttling over these valves causes a substantial loss of efficiency. One option for containing these losses is variable-pressure operation of the boiler.

Cycling also imposes added stresses on the steam turbine. A steam-bypass system is one way to reduce these stresses. Objective of the bypass system is to match steam temperatures with the metal temperatures of the turbine rotor and casings—avoiding large thermal stresses that occur if steam is appreciably hotter or colder than the turbine—by allowing the boiler to reach operating conditions before the turbine is rolled. In the process, superheater-outlet headers and the main steam piping are heated up before steam is admitted to the turbine, the secondary superheater is isolated to give the desired steam pressure and temperature, and feedwater flow and drum swell are stabilized during the restart sequence.

Varying levels of bypass systems are available. The simplest is a pipe and shutoff valve which exhausts a few percent of full-load steam from the boiler to the condenser. The steam can be drawn from between the drum and primary superheater, from between the primary and secondary superheater, and from between the boiler outlet and turbine inlet. Alternately, a pressure-reducing valve can be used between the primary and secondary superheater.

Options for containing throttling losses over turbine valves during part-load operation include: constant-pressure operation using the control valves over the entire load range, pure sliding-pressure operation with the control valves wide open over the load range, one-valve hybrid operation in which one valve is throttled before sliding pressure is used to reduce load further, and two-valve hybrid operation in which two control valves are closed before dropping pressure to reduce load.

SECTION III:
CONTROL THEORY

Chapter six:
General control
philosophy

Control systems are normally developed, for complex and/or unusual systems, after a dynamic analysis of the system to be controlled has been made. Performing a dynamic analysis allows comparison of various control schemes, since the analysis will indicate the need for certain control modes.

Analysis begins with a block flow diagram of the system to be controlled, including its inputs and outputs. Inputs to the system have effects on the outputs from the system. They can be the system variables which are adjusted by the controller to control the output variables. They enter the block where they are transferred into outputs. The transfer which takes place in the block is mathematically represented by a transfer function.

In Figure 1 a simple block flow diagram with two inputs and one output represents a continuous hot-water heating system.

In this arrangement hot-water temperature is controlled at a desired value or set point by adjusting or manipulating the natural-gas rate to the burner. There are two different ways of controlling this system: by input control or by output control.

Feedforward control

The simplest way to control the operation shown in Figure 1 is to control the natural-gas feed rate. In this type of control system the inputs are controlled with no regard for the values of the output variables. Figure 2 shows the simplified heating system being controlled by feedforward control.

The controller measures both the gas flow rate and the cold-water rate and

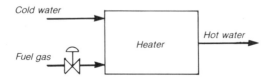

Fig 1: Simplified hot-water heating system

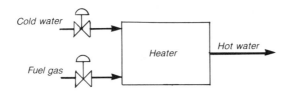

Fig 2: Hot-water system with feedforward control

compares them with their set points established to meet the desired water temperature. The advantage of this type of control scheme is that time is not lost waiting to see a change in the hot-water temperature. With any deviation in measured flow rates the control valves open or close, resulting in very fast response time. A slightly different feedforward control system may have the gas flow rate adjusted by the cold-water temperature and flow rate. Although the gas flow is controlled by deviations in cold-water temperatures it is still feedforward control because an input variable is adjusted by deviations in another input variable. This system works well when the desired outlet conditions remain constant. When the desired outlet conditions vary it is necessary to adjust the input variables to control the outlet variables at their desired values.

Feedback control

In Figure 3 this same heating system is shown operating with feedback or closed-loop control.

In this case the outlet temperature is measured and compared against the desired set-point valve. Adjustments are made by the controller to the gas

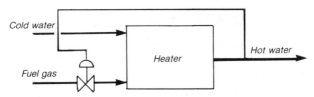

Fig 3: Hot-water system with feedback control

flow rate to meet the desired hot-water temperature. Lag time between taking response action on the manipulated variable and achieving the desired result in the controlled variable is the important difference between this system and the feedforward control system. In the feedforward system corrective action would have taken place before a deviation in hot-water temperature is measured.

In reality there are many more inputs and outputs to consider. For example, in steam-boiler control, fuel rate and heat content along with feedwater rate and air flow are some of the inputs that are considered in controlling steam pressure. Very often a combination of feedforward and feedback control is utilized in an integrated system.

Control action

In a feedback or integrated feedforward/feedback system there are many types of control action that can be taken to respond to deviations in measured variables. They range from simple on/off control to the controller anticipating how the deviation is occuring some time before making an adjustment.

On/off control

On/off control, the least expensive, most commonly used control method, is found in domestic heating and water systems, refrigerators, etc. When the measured variable is above or below its set point the controller simply opens or closes the final control element. Due to mechanical friction, the controller operates within an interval, taking action just below the set point and stopping just above it. To reduce wear, the frequency of control operation can often be adjusted by changing the dead band. In the hot-water heating system shown in Figure 4, natural gas is fed to a burner to heat water in a tank.

With on/off control, as the water temperature in the tank deviates from its

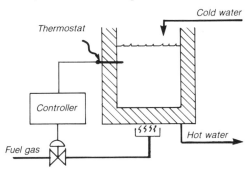

Fig 4: Hot-water control system

set point the gas valve operates within the dead band of the control element. The rate of temperature change in the tank is much greater when the burner is firing than when it is off. Therefore, the burner fires less than half the time. Assuming that heat is instantaneously transferred from the burner to the water, and that there is no flow of water, the hot-water heater would perform as shown in Figure 5a.

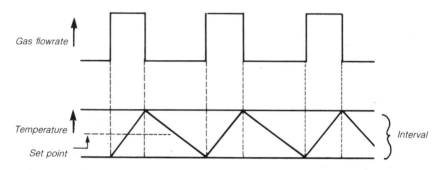

Fir 5a: Hot-water-system response to temperature, gas glowrate

While firing the temperature of the water rises above its set point; when the gas flow finally stops, the water temperature is somewhat above its set point. Then the temperature drops, at a slower rate than it rose during firing, until it travels through the entire interval, causing the gas valve to open and resume firing. However, heat is not instantaneously transferred to the water due to the heat capacity of the tank wall. Figure 5b shows the actual performance of on/off control of the hot-water heater.

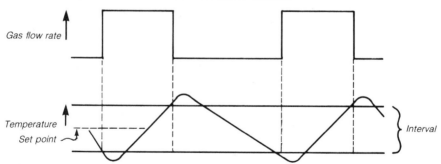

Fig 5b: Hot-water system response to on/off control

As the temperature of the hot water drops below the point where the controller turns on, the hot-water tank wall begins to heat. But before heat is transferred to the hot water the tank wall must be heated. This results in a further drop in water temperature. After the tank wall is heated it begins to transfer heat to the water until the measured hot-water temperature is at the upper level of the temperature interval. At this point the gas firing stops, but heat from the tank wall continues to heat the water before it begins to cool. Thus, the on/off cycle of control is greater than if there was no lag due to the heat capacity of the wall.

Proportional control

On/off control results in cycling within the interval set for the system. This cycling is not normally advisable in most control systems, especially steam-boiler systems. Modulating control as a function of the error eliminates this. When modulation is linearly related to the deviation of the output from its set point, it is called proportional control. This deviation of the output is the error signal perceived by the controller. Controller output varies with this error in linear fashion. As the change in this error increases the change in the controller output increases proportionally to the gain of the controller. For the hot-water system shown in Figure 4 temperature is measured and a signal is sent to the controller. This is compared with its set-point value. Deviations from the set point result in error signals. In proportional control, for a given error signal the controller sends a signal to the control valve telling it to move to a specific position. This valve position is dependent upon the proportional gain of the controller. Proportional gain 'K', the error signal 'E', and the controller output 'C' are related according to Equation 1.

$$C= KE \ . \ . \ . \ . \ . \ (1)$$

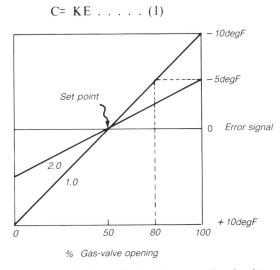

Fig 6: Fuel-gas control valve adjusted for proportional gain

Figure 6 illustrates how the gas control valve is adjusted for different values of proportional gain. As the measured temperature decreases from its set point a negative error signal is generated. When the error signal indicates that the temperature is 5 deg F below its set point, the gas valve opens 80% of full flow at a proportional gain of 1.0. When the proportional gain is 2.0 the control valve opens 100% in response to the controller-output signal. When a step change in feedwater temperature is applied the sudden drop in temperature results in a gradual drop in the hot-water temperature depending on the tank capacity and flow rate. Higher gas flow through the control valve results in response to the error signal generated. Eventually the water temperature in the tank stabilizes just below the set point. The error between this steady-state temperature and the set-point temperature is called the offset. Because more gas must be fired to heat the colder feedwater temperature this offset is unavoidable. Response to the step change is shown in Figure 7. If gain is high enough to cause overshoot, controller output is increased until the temperature of the water in the tank begins to increase. As water temperature rises, the error signal reduces, causing a reduction in output. This continues until the temperature starts to fall again. Several oscillations may occur before the final water temperature in the tank stabilizes.

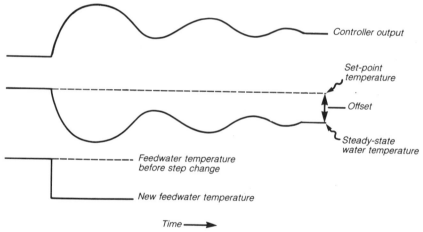

Fig 7: Feedwater temperature response to step change

Proportional action is taken by the controller when the measured variable is at other than the set point. However, the proportional action takes place within a given range of the measured variable. For example, with a proportional gain of 1.0 the controller would make adjustments between a fully open and fully closed valve for deviations of ± 10 deg F from the set-point value as shown in Figure 6. If the deviations from the set point are greater than 10 deg F the control valve is simply open or closed. An error band of 20 deg F causes the controller to vary its output by 100%. This range of error is known as the proportional band of the proportional controller. If the recorder chart of the controller were capable of measuring 100-deg-F variation in hot-water temperature, then the bandwidth would be 20% because it could be adjusted to measure 100 deg F of error. This bandwidth may be adjusted to minimize the offset experienced in load changes.

Integral control

In proportional control there is only one position for the control valve to assume for each value of the controlled variable within the controller's proportional band. Proportional action by the controller results in a new valve position which is proportional to the deviation from the set point. With an increased load placed on the system this new valve position does not permit sufficient gas to pass through the control valve to heat the water in the tank to its set point. As a result the tank-water temperature reaches a steady-state value below the set point. The offset between the set point and steady-state temperatures can be eliminated by integral control. Automatically resetting the proportional band changes the valve position that would have been assumed for a particular error. It is possible, by modifying the output, to have the valve open further for a given deviation from the set point. This permits more gas through the control valve, thus enabling the controller to maintain the water-tank temperature at its set point with no offset. When the load changes the output changes again in the opposite direction of the load, allowing the tank temperature to return to its desired value.

Output of the controller is proportional to the integral of the error for integral control. Equation 2 describes how the output varies, where Ti is the integral time setting, and t is time.

$$C = \frac{1}{Ti} \int_0^t E \, dt \ . \ . \ . \ . \ . \ (2)$$

The controller output is proportional to the integration of the error during the time the deviation of the controlled variable from its set point occurs. While integral control eliminates offset its initial response to deviation is

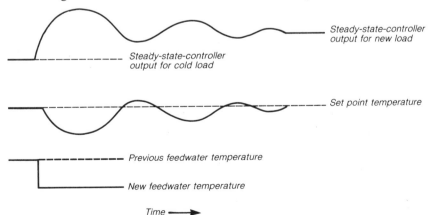

Steady-state-controller
output for new load

Steady-state-controller
output for cold load

Set point temperature

Previous feedwater temperature

New feedwater temperature

Time ⟶

Fig 8: Proportional and integral control response to feedwater temperature drop

slow. For this reason it is most often coupled with proportional control. Offsets created by proportional control are corrected for by integral control, which functions like a manual reset where the set point is adjusted after each load change. When these two control actions are combined they can be represented by Equation 3, where TR is the reset time or the time necessary for C, the controller output, to be the same due to integral-action control as it is for proportional after a step change.

$$C = K \left(E + \frac{1}{TR} \int_0^t E \, dt \right) \quad . \quad . \quad . \quad . \quad (3)$$

An illustration of how proportional and integral control respond to the step drop in feedwater temperature to the hot-water tank is contained in Figure 8.

Derivative control

Although the addition of integral control to proportional control permits the hot-water temperature to be controlled at its set point there will be temporary deviations until the reset feature provides the proper output. To minimize this, derivative control or rate action, which permits adjustments based upon the rate of change of the temperature from its set point, is added to proportional control. Taking an initially large corrective action when the error occurs makes the control valve open further than it would have with just proportional and integral control. After making this initial change it operates to remove its effect, while the other controls respond to determine the final position of the control valve. Adding this type of control makes it possible to minimize the lag time inherent in the other types of control.

Derivative action is directly proportional to the rate of change of the controlled variable or the rate of change of the error. Controller output in a system using proportional plus integral and derivative control would be represented by Equation 4, where TD is derivative time or the time interval by which the controller action is advanced above proportional control action. It

$$C = K \left(E + \frac{1}{TR} \int_0^t E \, dt + TD \frac{dE}{dt} \right) \quad . \quad . \quad . \quad (4)$$

is the time required for control-valve movement to its corrected position using proportional control minus the time required to reach that position using the combined control.

The addition of derivative control is very useful in startup. The initial overcorrection by the derivative-control action brings the controlled variable very close to its set point in a short period of time, thereby minimizing the effects of time lags. The controlled variable reaches its set point more quickly than with proportional plus integral control only.

news release
FROM LESLIE CO.

x

PLE

For Release: Immedia

Contact: Gene DeAngelo
Leslie Co.
(201) 887-9000

FIRST COMPLETE GUIDE TO MANAGING STEAM AVAILABLE

A new book entitled, "Managing Steam, An Engineering Guide To Commercial, Industrial And Utility Systems" is now available from Leslie Co., Parsippany, NJ.

The only complete guide offered on the subject, the 224-page volume contains major sections on: industrial and utility steam

Systems

FILE

applications. Detailed charts, schematics, diagrams and photos illustrate processes throughout the guide.

Leslie Co. is a leading supplier of high-performance steam management systems and services for industrial, commercial and marine markets.

For more information contact: Leslie Co., 399 Jefferson Road, Parsippany, NJ 07054, (201) 887-9000.

# # #

LES-3043
RNEC

$35 per
Hemisphere Co
Publishers
p 98

Parsippany, New Jersey 07054   Telex 136445 • Telephone (201) 887-9000

Combined feedforward/feedback control

While feedback control has the advantage of making adjustments to manipulated variables based upon the deviation of the controlled variable from its setpoint, it has the inherent disadvantage of slower response to load changes than feedforward control. Feedforward has the disadvantage of being unresponsive to load changes and requires measurement of all inputs, some of which are difficult. Thus when there is a substantial lag time in measuring the system's output it may be preferable to combine both feedforward and feedback control. With a combined system load changes can be handled to return the controlled variable to its set point with very fast initial response.

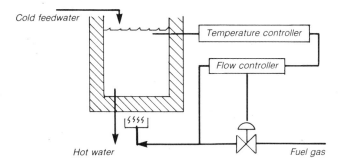

Fig 9: Combined feedwater/feedback-control system

A simple example of combined control is shown in Fig 9. Feedforward control is applied by controlling the flow rate of gas to the burner for a given load. The temperature in the tank is measured and compared to its set-point value. When there is a deviation in temperature from its setpoint the set point of gas flow rate is readjusted and feedforward control maintains a constant flow of gas to the burner.

Control-system stability

To maintain stable system control the overall controller must be set below a maximum value. Above this maximum value, response to load change results in undamped oscillations away from the desired output, with the error growing with each cycle.

Maximum gain can be determined by using frequency response to analyze a control system. Frequency response is the response of the system to sinusoidal inputs covering a wide range of frequencies. Maximum gain for stable operation and the critical frequency of the system is obtained from this analysis.

These values are obtained from open-loop frequency response, the response of the system with the control loop broken at some point, usually after the controller. In this case the input is a sinusoidal signal to the control valve, with the output being either the controller signal or process output. Since gain is unknown before performing this analysis, process output is used for frequency-response plots. Bode diagrams are used to present the response data at various frequencies.

The reponse of a first-order system to a sinusoidal input consists of both steady-state and transient terms. Within a few cycles the transient term disappears, leaving the steady-state response which is a sinusoidal wave of the same frequency as the input. This wave can be characterized by the ratio of the amplitudes of the output to the input or the amplitude ratio and the phase angle. The phase angle is obtained from the phase lag. In a system where frequency is low, output is nearly equal to input and the result yields a very small lag. Conversely, high-frequency results in much lower output and large output lags.

Bode plots consist of log-log plots of amplitude ratios versus frequency and semi-log plots of the phase lag verus frequency. Maximum gain is the reciprocal of the amplitude ratio at the frequency where the phase lag is 180 degrees.

Chapter seven:
Boiler control

The main objectives in the control of a steam boiler are to provide steam of desired quality and quantity generated as economically as possible. Any one of these objectives can be met with fairly simple controls. However, to meet all of these objectives an integrated control system is necessary. Control consists of responding to changes in input variables to maintain output requirements. Response is accomplished by adjusting the inputs to control the output steam conditions of flow, pressure, and temperature at the desired values or set points within preset tolerances. In controlling the boiler operation to meet the output objectives the actual values of pressure and sometimes temperature, depending on superheat, are measured and compared with the set-point values for each of these output variables. Deviations from the set points require corrective action on the part of the controller. The type of action to be taken depends on the degree of complexity of the control system. A control system can be either manually or automatically controlled but the basic parameters which can be adjusted to correct the output steam conditions are fuel, air, liquid level, or feedwater flow rate depending on the type of boiler. How these three parameters are varied to control the output is the heart of boiler-control philosophy.

Control philosophy can be simple or complex and it can involve manual or automatic control. Boiler operation can be performed by manual control but this is not practical in most cases due to the frequency and speed with which control action must be taken.

Control action must be taken whenever the output value does not match the set-point value. Output conditions vary from their set points due to disturbances which can be categorized as changes in steam demand, changes in fuel-heating value, and cycle efficiency. Steam demand in an industrial complex or power plant varies because steam demand of individual users varies or electric-power demand changes. Depending on the nature of the disturbances a variety of control schemes can be implemented to adequately control the output conditions of the steam produced.

Control systems act upon the fuel, air, and feedwater rates. Feedwater control has the objective of maintaining a constant steam-drum water level. Combustion controls to regulate air and fuel flows to the boiler furnace must meet the objectives of producing adequate steam efficiently.

Feedwater controls

For drum-type boilers feedwater control is used to maintain the water level in the steam drum within the desired limits. Water-level changes in the drum are caused by changes in steam flow rate, water flow rate, steam-drum pressure, water temperature, and the rate at which heat is added. There are three types of system employed to control the water level in a drum: one-, two- and three-element control.

One-element control

A steam drum typically is fitted with a level transmitter which sends a signal to the feedwater controller where it is compared to the set point. Corrective action is then taken by the controller by sending a signal to the feedwater control valve. Normally control action is of the proportional plus integral type.

As steam load on the boiler increases it would be expected that drum water level would drop, resulting in a response by the feedwater control system to raise the feedwater rate. However, as this increased load is placed on the system steam-drum pressure drops, causing the steam bubbles in the drum to increase in volume. This leads to a reduction in the density of the liquid, thus raising or swelling the drum water level. While the level is higher in the drum the density of water in the drum is lower. Likewise, in a reduced-load situation drum pressure increases. This causes steam bubbles to reduce in size resulting in shrinkage of the water level. A lower water level causes the one-element controller to admit more feedwater to compensate for the lower level. In both increased- and decreased-load cases the single-element controller initially responds in the wrong direction.

In the case of a higher steam load and an increasing drum level the rate of steam removal is greater than the feedwater rate. This causes even more steam bubbles to swell, causing level to rise. However, the actual mass of water in the drum is falling as this is occurring. Soon the water level drops quite rapidly. Level control then responds by admitting more water. Feedwater, being relatively cool, will collapse steam bubbles below the surface, again reducing the drum level. As a result, whenever large load changes are applied a cycle is experienced in the feedwater rate to the boiler. This also upsets the heat balance of the entire boiler and the feedwater heating system. Whenever large load changes are applied it takes a long time to restore the drum level to its set point. For this reason, one-element control systems are restricted to boilers which have relatively large drums or which experience relatively slow load changes.

Two-element control

Effects of shrink or swell are felt by the steam drum immediately following a load change. Their effects are reversed as the transient works its way through the system. When an increased steam load is placed on the system the drum level eventually drops, but not until after the incorrect action has taken place. Outlet steam flow can be utilized to correct for this effect. When an increased load causes steam pressure to drop and liquid level to rise, the steam flow can be used to adjust the feedwater flow rate in a feedforward type of control. Steam flow rate is used as an input variable to the system to control or adjust the feedwater rate to the drum. Feedwater is controlled in direct proportion to the steam flow measured. As this control is being taken the drum-level controller takes proportional action on the error between the set point and the level. Valve position is determined by the sum of the signals from the drum-level element and the steam-flow element.

Three-element control

In the two-element control system the controller output was determined based upon both steam-flow and level measurements. Steam flow measured resulted in a feedwater demand which would match that leaving as steam, while the level measured in the drum resulted in a reduced feedwater demand due to swell in the increased-load case. The sum of these two resulted in the combined signal that made the controller position the control valve.

A feedforward control system utilizing three elements maintains feedwater flow equal to the feedwater demand. As described above, the feedwater demand is determined by the summation of the signals from the level and steam-flow elements. Feedwater is measured and compared to the feedwater demand signal; the difference between these is the controller output. Valve positioning is regulated by proportional plus integral action.

Systems employing three-element control can be adjusted to return drum level under any loads, or in cases of wide load changes the system can be adjusted so that the water level varies with load to compensate for shrink and swell. Three-element systems are seen in large boilers since cost considerations dictate small drum sizes compared with flow rate. This system is also well-suited to smaller boilers which see wide and rapid load changes.

Feedwater-control selection

There are many items to consider in the selection of a feedwater-control system; general guidelines for the selection of a system are available. Considerations include the type of boiler, drum-solids concentration, steam loads, and feedwater-pressure control.

Combustion controls

Fuel-firing rate must be adjusted with variations in steam-load demands. Variations in the steam pressure at the outlet are usually indicative of imbalance in the boiler-heat balance. Thus, the fuel-energy input is regulated to match the energy withdrawn by the output steam.

Steam pressure is measured and compared to its set-point value by the boiler-pressure controller which produces a firing-rate signal. After this signal is transmitted to a boiler master both fuel and air are simultaneously adjusted to satisfy the energy demand in order to maintain steam-header pressure.

In like fashion, removal of combustion products is controlled by furnace pressure. When furnace pressure rises there is either excess forced draft or a deficiency of induced draft. As the furnace pressure falls the opposite is true. Regulation which is responsive to pressure variations makes adjustments to the uptake damper or forced-draft fan to maintain desired pressure. To ensure that the fuel is combusted in the most efficient manner excess air is minimized.

Fuel flow is measured by orifice metering for oil and gas; stoker speed for solid fuels and a variety of other methods for pulverized-coal boilers. Coal-flow measurements include the use of the speed of a raw-coal feeder with constant volume displacement, the coal-carrying capacity of air, and a measurement of the power requirements for grinding the coal.

Combustion controls can be categorized into two classes of system, parallel and series. Parallel-system control uses the steam pressure to adjust fuel feed and air flow. Series systems make adjustments to these variables in three different ways. In the first system the steam error signal adjusts fuel flow. A measure of this flow rate is used to set the air-flow rate. Alternatively, in a similar series arrangement the air flow and fuel flow are interchanged. The first case, where air flow is set by fuel flow, is better known as air flow following fuel flow. This system will not permit air flow to exceed that which is required for the maximum possible fuel-firing rate. In the fuel-following-air system any failure in air flow will automatically restrict the fuel flow to that corresponding to the available air supply. A third system has both fuel flow and air flow adjusted based upon signals generated as a result of steam pressure and steam flow. As in the first series arrangement, fluctuations in steam pressure result in fuel-flow corrections to maintain steam pressure. Steam output from the boiler is utilized to estimate the required rate of heat input to the furnace. Steam-flow measurement provides a signal for regulation of air flow. When fuel-flow measurement is not reliable this system is preferred.

Startup and low-load control

During startup it is necessary to fire at low heat input for extended periods of time to prevent, among other things, overheating of superheaters and reheaters. Throughout the initial period combustion efficiency is poor, creating substantial amounts of unburned hydrocarbons.

Because of its low ignition temperature gas is an excellent choice as a fuel for startup. It is important to prevent any momentary interruption in the flame during startup. Although fuel oil has good ignition stability it has other problems that make it less attractive as startup fuel. After a relatively short period of time the fire will burn clearly and low-pressure boilers will be prepared to take the load. While the fire burns clear, the furnace temperature is still low and the stack will discharge a dark grey smoke. Accumulation of significant deposits of oil or soot—more common when using fuel oil—can be seen on the air-heater and economizer surfaces. If these deposits accumulate to a significant degree there is increased potential for a fire since the deposits ignite easily. Sootblowing is used very often to continuously blow steam across the surfaces of regenerated-type air heaters, while the sootblowers rotate in the gas flue. Although gas startup is preferred, fuel-oil startup is sometimes necessary. Light fuel oil with steam or air atomization is most desirable, heavy fuel oil with steam atomization is second best, and heavy fuel oil and air atomization are the least preferred method of startup. Startup with oil being atomized by mechanical atomizers is normally not considered due to the oil and soot deposits on the economizers and air heaters. Boilers having mechanical atomizers for operation at 20% of load or greater are started up on gas or oil as described above.

Steam demand varies greatly. Therefore, feedwater, fuel, and combustion air flows must be adjusted to replace the water and heat removed by the steam. Flow signals in most boiler systems are linearized due to the wide variations in output and to the need to keep the gain constant over the range of operation.

Safety

A potential hazard always exists when burning any fuel. Gas diffuses very rapidly. Therefore, any leak into an enclosed space may result in an explosive mixture of air and gas. Oil is normally not in a form where it could disperse; moreover, it has a higher ignition temperature than gas. Nevertheless, it, too, has the potential for forming explosive mixtures when its volatile components vaporize and mix with air. To ensure safe operation these mixtures should not be permitted outside the combustion chamber and only when the burner is lighted. For this reason the furnace is purged with air or nitrogen before any light or spark is introduced or before relighting after all flame is out. Burners can be ignited without purging only if another burner is already in service. When igniting a burner the light or spark-producing device must be operating before any fuel is supplied to the burner. To prevent backfire or flareback of the flame the location of the igniting device must be carefully chosen and the device must provide a continuous source of ignition until stable operation is obtained. Typical safety controls to ensure proper operation after startup provide protection from flame failure, low fuel pressure, loss of combustion air supply, loss of induced draft, low feedwater level, and excessive steam pressure. Flame failure is prevented by incorporating flame sensors to monitor the igniter flame during lightoff and the main burner flame during normal operation. If the igniter fails to light during startup within a safe period the controls will close fuel valves and require manual reset and firebox purge before resuming normal operation. In each of the other cases mentioned above the shutoff of fuel is provided for to prevent an unsafe condition.

Control for power generation

As previously described, control action is taken whenever output does not match the set point. Steam demand in an industrial complex is dependent on a maze of individual users. Each one's steam demand can fluctuate widely. In an electric powerplant steam-demand changes are tied to variations in electric-power demand from the turbine/generator which is controlled by its throttle (or inlet) pressure. Three types of control system for electric-power production are turbine-following, boiler-following, and integrated boiler-turbine/generator control.

Throttle-pressure control

For normal steady-state operation the turbine control valves are manipulated to maintain a constant pressure at the turbine inlet. A steady-state throttle pressure program takes megawatt demand into account in establishing the set point for steady-state throttle pressure. Turbine control valves will maintain constant throttle-pressure during steady-state operation.

Actual throttle-pressure is measured and compared to this normal set point, resulting in an error signal being generated whenever the measured value does not match the set point. Throttle-pressure control will make an adjustment based on this error to the position of the turbine control valves in order to return the throttle pressure to its set point. When a throttle-pressure error is generated a signal which is proportional to the error changes boiler demand to bring pressure back to its normal value.

Steady-state throttle-pressure set-point values are modified during electric-power load-demand changes. The objective of the steady-state throttle-pressure program is to program the turbine control valves in such a manner as to provide fast response to megawatt load-demand changes and maintain stable throttle-pressure control at the same time.

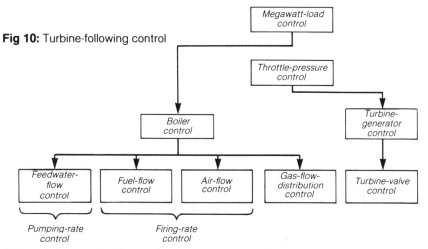

Fig 10: Turbine-following control

Turbine-following control

In this configuration responsibility for throttle-pressure control falls on the turbine/generator while megawatt load control is entrusted to the boiler. As a load-demand increase is placed upon the system the boiler control will respond by increasing feedwater and heat input. This results in an increase in throttle pressure. To control throttle pressure at its set point the turbine control valves change position to admit the increased boiler output, thereby increasing load. Reduced loads on the system result in control that will reduce throttle pressure before turbine controls close the control valves to reduce load (Fig 10).

A basic disadvantage of this scheme is slow response. Before the turbine control valves reposition to make a change in load, the boiler must change its output by adjusting feedwater and heat input.

Boiler-following control

In the boiler-following configuration responsibility for throttle-pressure control falls on the boiler while megawatt load control becomes the responsibil-

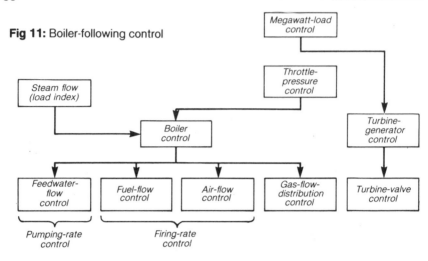

Fig 11: Boiler-following control

ity of the turbine/generator. When a load-demand change occurs the turbine control valves are repositioned so that the turbine/generator can meet the new load. After this load change the boiler modifies the feedwater and firing rates to achieve the new load and eventually return the throttle pressure to its normal set-point value. In this system response to the load change is fast, thanks to its ability to take advantage of the energy stored in the boiler. But this fast response brings with it less stable throttle-pressure control since it must wait for the boiler to meet the new load conditions (Fig 11).

Integrated boiler-turbine/generator control

As we have seen, both of the above mentioned control schemes have advantages and disadvantages. By combining the two into an integrated system, it is possible to exploit the best features of both systems (Fig 12).

In the integrated system responsibility for throttle pressure and megawatt load control is accepted by both the boiler and the turbine/generator. The system takes advantage of the stability provided by turbine-following control. Use of the stored energy in the boiler provides for fast response to load changes. Initial response to load changes is provided by the turbine/generator because the boiler is unable to produce the steam required at constant pressure rapidly enough. As the load change is applied, the throttle-pressure set point is adjusted based on the megawatt error. The turbine control valves are repositioned to respond to the new set point and produce the new megawatt load very rapidly. Throttle pressure is restored as the boiler adjusts to achieve the new load. As this occurs the throttle-pressure set point returns to its normal value. Assigning the control responsibilities properly to the boiler and turbine/generator allows electricity to be produced efficiently and

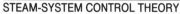

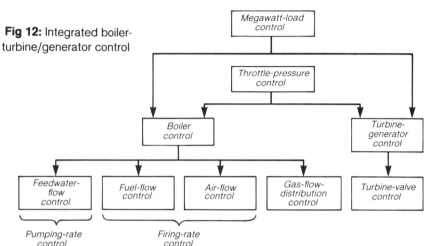

Fig 12: Integrated boiler-turbine/generator control

changing load demands to be met rapidly. Response time is substantially greater than that obtained in the turbine-following system but not quite as fast as that achieved in the boiler-following configuration. This is because a balance must be maintained between boiler response time and stability.

Integrated boiler-turbine/generator control is effected by ratio control of certain inputs. Related inputs are controlled in certain ratios. These are boiler energy input to generator output, fuel flow to feedwater flow and to air flow, recirculated-gas flow to air flow, and superheater-spray-water flow to feedwater flow. Megawatt load demand leads a parallel control system of turbine/generator and boiler control which makes use of ratio controls. The area-load control system utilized by the electric utility imparts a demand signal to the boiler-turbine/generator unit. Megawatt load control evaluates these demands on the basis of the units' ability to respond to load changes, then it instructs the system to make the changes required to meet the load change. After evaluation of load limits and maximum rate of load change the megawatt load demand is adjusted for any frequency error. When the frequency error occurs the turbine-speed governor changes the turbine-control-valve position, thereby changing load.

Integrated boiler-turbine/generator load control is used mostly by large utility companies to economically dispatch their units.

Comparison of conventional boilers

Steam-boiler types include natural-circulation, controlled-circulation, combinations of both, once-through, recirculating, subcritical, and supercritical. In natural-circulation boilers circulation is produced based on the difference in density between water and steam. Circulation is improved with the addition of a pump in a controlled boiler. Different conventional boilers are described below.

ONCE-THROUGH-BOILER CONTROL

Once-through boilers operate at full steam temperature over wider load ranges and can operate at supercritical pressures; this gives them increased cycle efficiency. In a once-through boiler the steam output is dependent only on the heat input via feedwater. Outlet conditions are a function of the amount of heat input and the flow rate of feedwater. Pressure level may be varied by an outlet control valve. When this pressure is held constant, the temperature of the outlet steam depends only on the heat input. In normal steady-state operation feedwater flow is controlled to match steam flow. Outlet steam pressure is dependent not only on the position of the control valve but also on the fluid density within the system. Thus, changes in heat input to the system will affect steam pressure and temperature. Therefore, it is possible to control steam flow and pressure by outlet-valve positioning as well as by controlling the feedwater flow and heat input. If the feedwater flow is changed without changing the heat input the outlet steam temperature will change.

In a once-through boiler water is pumped into the unit as liquid. It passes sequentially through all pressure-part heating surfaces, where it absorbs heat and is converted to steam at the desired temperature. Water is not recirculated in the unit and therefore a drum is not required to separate the steam from water. Steam flow may be controlled by feedpump speed, with steam pressure being controlled by turbine throttle.

A minimum flow must be maintained inside the furnace circuits to prevent overheating of furnace tubes. Before startup, this circulation must be established at minimum design flow through all pressure parts that are exposed to elevated temperatures.

RECIRCULATING BOILERS

Forced circulation is provided by pumping the feedwater through the heat-absorbing circuits in a recirculating-type boiler. While all of the water in the once-through boiler evaporates to steam, in the recirculating boiler the amount of water recirculated is in considerable excess of the steam produced. A drum is required to separate the steam and water. This separated water is combined with the feedwater before being recirculated through the heating circuit.

Since the volume of feedwater added to the recirculating water is relatively small, the amount of cooling lowers the water temperature only slightly below the saturation temperature. As a result, the entire waterwall operates at essentially constant temperature. This uniform temperature does not exist in the once-through boiler, where the waterwalls are exposed to high rates of heat absorption.

Circulation in a once-through boiler is of paramount importance any time the boiler is fired to prevent burnout of exchange surface. In the drum-type boiler this is not a problem since recirculation adequately meets this need even during startup.

The types of controls used to maintain desired steam flow and conditions are the same as those described in the section on feedwater controls for boilers. One important parameter controlled is the drum water level.

SUPERCRITICAL BOILERS

Like the once-through boiler the supercritical does not require a steam-water separator (drum). This means that provision must be made for performing certain functions of the drum. For example, there can be no blowdown control, so condensate treatment or polishing is necessary. Also, a boiler extraction valve may be located ahead of the superheater to reduce the firing rate required during cold startups. This also permits a hot quick restart without excessive chilling of the high-temperature portions. As in the once-through boiler, a minimum circulation of water must be maintained by a feedwater pump to avoid overheating exchange surface.

NUCLEAR STEAM GENERATORS

The integrated boiler-turbine/generator control system is used to control the production of power from fossil-fuel-fired plants by turbine control and boiler control. An integrated control system for nuclear-fueled plants includes turbine, steam-generator, reactor-coolant-loop, and reactor controls.

A nuclear steam supply used to generate electric power consists of a nuclear reactor, pressurizer, steam generator (in PWR systems), and turbine/generator. Coolant water is circulated through the reactor, heated and pressurized and pumped to the steam generator where its heat is used to generate steam before being reheated again in the reactor. Steam produced in the steam generator drives the turbine/generator to produce electricity.

Steam-generator controls are based on adjusting feedwater control to meet megawatt demand, as in a conventional fossil-fueled boiler. Turbine controls match output with megawatt demand. Reactor control and reactor-coolant-loop control are unique to the nuclear steam-supply system.

Control rods are used to regulate the heat release or power output of the reactor while maintaining a constant average reactor temperature. Average reactor temperature is summed with the megawatt demand in the power controller, resulting in a reactor-neutron-power demand. Power-controller output is a signal which is compared with actual neutron power measured by the power-range neutron detectors. An error generated results in an adjustment in the control-rod drive to either withdraw or insert to regulate heat generation. A control-rod program selects the appropriate control-rod-drive group sequentially. This is how normal control is effected (**Fig 13**).

When there is input from a protective function, control commands are interrupted and protective logic will carry out instructions according to the

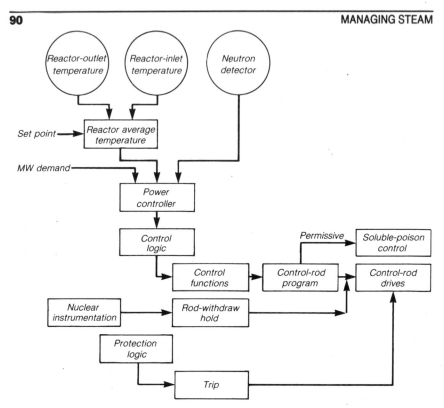

Fig 13: Nuclear-powerplant control scheme

input information. For example, an input which results in instructions to hold the rod-withdraw controls will prevent all control rods from withdrawing. The control-rod program will be bypassed and a signal will cause all rods to drop into the reactor core. Reactor controls maintain constant average reactor coolant temperature for loads of 15 to 100 percent with the steam system operating on constant pressure at all loads. Average reactor coolant temperature drops below 15 percent of load.

Average reactor coolant temperature, megawatt demand, and reactor power are input signals received by the reactor power controller. When an error signal is generated by the controller action is taken by the control-rod-mechanism position until error is corrected.

Power demand is calculated from megawatt demand, average reactor-coolant-temperature error, and time. This demand is compared with actual reactor power to generate a power-error signal. Control-rod-drive assembly action is taken when the error exceeds the dead-band settings.

Automatic control of reactor power level is assigned to a portion of the control rods which are usually arranged in three symmetrical groups operating sequentially. An automatic-sequence logic unit which permits operation of no more than one group at a time is employed.

When the reactor is being started up, control rods are withdrawn from the core in symmetrical groups in a preset sequence. The group size is also preset and a control-rod grouping panel is used to make individual rod assignments.

Reactor coolant controls are designed to maintain coolant pressure constant as well as level in the pressurizer. These controls act outside of the integrated control. To maintain constant coolant pressure, the pressure controller calls for the pressurizer heaters to come on as coolant pressure drops, and for the pressurizer spray valve to open when coolant pressure rises. Level control in the pressurizer is simply maintained by trimming the makeup-line control valve. Pressurizer-heater power is cut out when low level is detected in the pressurizer (Fig 14).

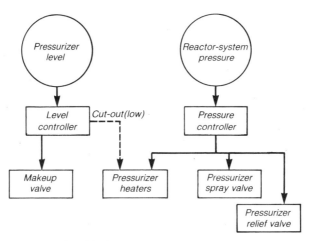

Fig 14: Pressurizer-system control

Steam-generator control matches feedwater flow to megawatt demand using steam-generator outlet pressure as the controlled variable. Megawatt demand is converted to a feedwater demand and the outlet-pressure error is also converted to a feedwater demand. Both demands are summed to get total feedwater demand. Actual feedwater flow is compared to this demand and the resulting error is used to position the feedwater control valve.

Steam-generator water-level control can be used when operation falls below 15 percent of load. Turbine-bypass valves limit steam-pressure rise at operation below 15 percent of load. Additional steam-generator controls are included to limit the loads to those the turbine or steam/generator are capable of handling; to limit water levels to prevent overpumping of feedwater; to limit feedwater demand to flows which the pumps have capacity for or when reactor outlet temperature or feedwater temperatures are low; and to limit the megawatt demand.

WASTE-FIRED BOILERS

Waste fuels are by-products which are relatively inexpensive as compared to conventional fuels. In designing a control system, consideration must be given to the type of fuel to be burned. Special concerns not normally encountered with conventional fuels include variability of heating value, fuel handling, flow measurement, and fuel properties such as toxicity, corrosiveness, noxiousness, water content, ash content, etc. Typical waste-fueled boilers are base loaded to some level with the waste fuel. Load variations are handled by adjusting the prime fuel, which is usually gas or oil. As the prime fuel takes most of the swing and is the most expensive fuel its consumption should be minimized. Likewise, overall boiler efficiency should be maximized.

In order to minimize consumption of the prime fuel it is necessary to manipulate the waste fuel to control steam pressure. Problems which are encountered in addition to the ones mentioned above are dynamic response and combustion stability.

Identification of all items relative to the controlability will enable a system to be designed which, though more difficult to control than a conventional system, will still be manageable.

Four methods can be used when firing waste fuels:
-firing of waste fuel only
-baseloading of waste/adjust prime
-baseloading of prime/adjust waste
-control-ratio waste/prime

Firing of waste fuel only can be accomplished if prime fuel is not needed to support combustion and if the dynamic response of the boiler can meet steam demands. Also, some reliable means of measuring fuel must be developed. Inferential measurement is usually necessary and may include the use of boiler heat balance, fuel properties, or belt-scale weighers.

Baseloading of the waste fuel depends on the capacity required from the boiler, which may limit its use. Baseload for the waste fuel can be set manually but should be limited to the actual firing-rate demand in the event of a large drop in steam demand. Prime fuel handles increases in steam-load demands.

Prime fuels may be baseloaded when a constant ignition-stabilizing source is required and also when the waste-fuel firing rate must be limited. Baseload is determined by the rate of-change limit and steam load pattern. For rate limiting of both increasing and decreasing demand the baseload must be at a value which permits the prime-fuel firing rate to be decreased sufficiently during transients. Baseloading of prime fuels is most often used in cases where the waste-fuel rate is limited for increasing demand only. In this case the baseload for the prime fuel is that required to maintain stable firing.

Maintaining a fixed ratio of waste to prime fuel may be required to maintain stable combustion. A multiplier is used to determine the proportion of total firing-rate demand that is transmitted to the waste-fuel feed. Then at steady state a subtractor calculates the net prime-fuel demand by subtracting the waste-fuel demand from the total demand.

Optimizing boilers for load allocation

In industrial powerplants the "swing" boilers, which are equally located, handle demand variations. By allocating loads among parallel boilers operating cost can be reduced.

A process optimizer is used to allocate the steam load among the total number of parallel boilers. Its function is to minimize the total operating cost of these boilers. Boiler-efficiency curves give characteristic efficiency equations. These are used to calculate boiler operating cost by the optimizer. The sum of the operating costs for all the parallel boilers in operation is minimized by the optimizer and the boilers are operated at their individual minimum costs.

As part of the optimizing system several software packages are employed. Calculation involves data acquisition and calculation. Handling supervises optimization and control, deciding when to optimize. Operator Interface imparts constraints to limit or input the range of key variables such as fuel cost, electrical cost, boiler status, and boiler reserve capacity. Boiler Control functions as either a supervisory or direct digital control for the implementation of actual load allocation. Optimization determines the load each boiler will carry. It determines the total and individual operating costs. This system can be used in either an advisory or closed-loop form. An advisory system simply provides instructions while the closed loop automatically performs the actions that the instructions in the advisory system would outline.

Chapter eight:
Steam-turbine
controls

Three basic controls are used for steam turbines. They are governor speed control, overspeed trip control and microprocessor control. Overspeed-trip control comes into play when governor speed control fails. Microprocessor control is a master control that can control the other two.

Governor speed control

According to American Petroleum Institute (API) specification No. 612 a governor speed-control system is designed to prevent the unit from exceeding the turbine trip speed by means of an instantaneous loss of rated load. Unless otherwise specified it shall control the turbine at all speeds within the specified range. Governors generally include mechanical-shaft, mechanical-hydraulic, electrical-hydraulic, and electrical-pneumatic types. Mechanical-shaft governors are not generally used today although a few systems still have them.

Mechanical-hydraulic governors are much more common, being used on all sizes of turbines. Oil-relay types are used mostly with single-stage turbines. While this governor will provide good speed control it is unable to prevent the turbine from overspeeding. Another major disadvantage of the system is the number of moving parts which are subject to wear. Since their hydraulics oil levels must be maintained; this just adds one more area for improper maintenance. Another type of mechanical-hydraulic governor has a trip and throttle valve, a speed governor, steam-admission nozzle valves, and a servomotor to operate the nozzle valves. Speed control for this governor is good. While response for closing of the admission valve is faster, in a situation where there is 100 percent loss of load it is not fast enough to control. It also has the disadvantage of many moving parts.

Electrical-hydraulic governors have the distinct advantage of fewer moving parts. One system has an actuator mounted on the main servomotor which responds to speed-governor drive signals. The actuator converts the signals to mechanical movement of the servomotor relay valve. Positioning of the turbine-steam nozzle valves is by this movement of the relay valve. Very good turbine speed control is obtained from this system and speed is controlled upon instantaneous loss of load. Speed control is established at about 1 to 2 percent of the initial set point. There is no mechanical overspeed trip but dual redundant overspeed trip relays set at about 5 percent of rated speed do provide overspeed trip control.

Another electrical-hydraulic system which incorporates some of the latest technology moves the steam-admission valves in a very different way. Electric speed-control signals are fed to a servoamplifier that operates servoactuators which move the valves thus controlling steam flow. Response time of these valves is very fast. When a changing-speed signal is sent, a very small increase, above the set point, in speed is seen at sudden load loss.

Overspeed trip control

According to API code, trip speed is the speed at which the independent emergency overspeed device operates to shut off steam to the turbine. It shall be approximately 110 percent of the maximum continuous speed. Two important considerations in the design of overspeed trip control are the speed which the rotor can take before damage and/or complete component part failure will occur; and the maximum speed the rotor will reach before it slows after the emergency overspeed device cuts off the turbine steam.

The speed a rotor can withstand depends on its type. Maximum speed can be affected by many factors but the key one which can be calculated is that based on time constants. They range from half a second up to ten seconds. When instantaneous loss of load is experienced and inlet steam is not shut off, acceleration to about 125 percent of rated speed will occur in a very short time, on the order of one- to two-tenths of a second for a time constant of one-half to one second. Even though the trip and throttle valves are activated at 110 percent of rated speed the turbine can reach more than 125 percent. This is because the typical trip and throttle valves take two- to three-tenths of a second to close.

In the process industry instantaneous loss of load does not occur as it would in the case of a turbine-driven generator. Here, where turbine-driven compressors and turbine-driven pumps may experience at worst a coupling failure, one to two seconds would pass before 100 percent full-load loss was sensed. Turbine-driven generators can lose full load in fractions of a second.

Mechanical overspeed trip mechanisms, while prevalent for many years, have some distinct disadvantages. These are: many moving and wearing parts, difficulties with overspeed-trip settings, inability to change settings while operating, need to have several mechanical linkages, and weight of the mechanical trip on the rotor. With electronic overspeed trip devices these disadvantages disappear but the system is, of course, totally dependent on its power supply. Uninterruptible power sources are used to overcome this main disadvantage.

Overspeed-trip systems must be independent of the governoring system. When both are electronic independent power sources should be used for each.

Microprocessor control

Microprocessor-based control systems can perform all the functions necessary for the control of turbines. Included in its range of capability are monitoring, automatic sequencing, load and speed control, maintenance of historical records, prediction of required maintenance, and safety and reliability. A microprocessor system can perform all of these tasks simultaneously and with much greater speed than is seen in separate overspeed trip or governor controls.

If sensors and actuators with a high degree of reliability are chosen carefully, this type of system will provide excellent control. When a sensing function is critical to operation the use of redundant sensors is indicated. Also, by prioritizing functions within the microprocessor itself it is possible to have multiple tasks performed by a sensor. Alarms, trips, interlocks, and remote indication can be handled with the same sensor.

In the design of actuators redundancy is again the keynote. Two actuators are of greatest importance, safety-trip and speed-control actuators. Solenoids are usually used to trip the unit for safety purposes, while relatively small electrohydraulic devices are used for speed control. Reliability of these actuators is of paramount importance because this is the final control.

SECTION IV:
HARDWARE FOR
STEAM-SYSTEM CONTROL

Chapter nine:
On hardware in
general

Implementing the control strategies discussed in the previous chapter requires the proper hardware—valves, controllers, sensors, regulators, actuators, packing and seals, metering devices, steam traps, strainers and others—specified to meet the control function at hand while offering long life and reliability in the environment it serves.

The valve is a good place to start. In the simplest sense, a valve is a mechanical device to shut off, open, or vary the flow of a fluid into, inside of, or out of vessels, piping, and tubing. Sometimes, controlling the flow of a fluid is intended to regulate another process variable, especially temperature, pressure, chemical concentration, and others. A valve opens or closes fully or partially to regulate this flow by moving some obstruction into and out of the flow path. Many components are part of the valve—packing, seals, actuator, bonnet, stem, trim, etc. These components, integrated into a highly engineered package, all combine to resist internal fluid pressure, temperature, dynamic effects, corrosion, erosion, and damage. Upon this basic function can be built a control system as elaborate as the system's designer wants or requires.

First of all, the valve's actuation can be performed automatically or manually. Automatic control obviously adds many components to the basic function. Signals are sent between the valve and some sensing device, sometimes initiated automatically by computer or by an operator from a control room. Pneumatic, electric motor, hydraulic, and electrohydraulic systems are available to automatically power a valve.

In many cases, an integral valve package, commonly called a regulator, can be purchased which includes many of the automatic control elements. These devices, for certain services, are comparable in performance to the components bought separately. However, as will be shown later, they have limitations.

A valve rarely operates independently. Rather, its function must be coordinated with a variety of other functions within the control system. Each function that's added also adds interdependencies which involve many design and operating tradeoffs. All of them must be considered before a valve and its auxiliaries are successfully matched to the particular service.

At first glance, the range of hardware is overwhelming to the uninitiated. But after the process parameters are established, the field becomes limited and more manageable very quickly. In gaining an understanding of what's available, the first task is to become acquainted with the basic valve components and the important design areas. Then, review the various categories of valves and their auxiliaries and what general duties they perform. Next, compare equipment that can function in a given situation for performance, reliability, and cost-effectiveness. Finally, review the special topics of valves and controls as they relate to the steam/condensate system. Some of these include: cavitation, noise control, fire safety, and control-equipment sizing and selection.

Here are some characteristics that apply to all valves (Fig 1).

The body, or shell, is the framework that holds everything together in a valve. It must resist fluid-pressure loads and loads from connected piping, support an actuator, the device that moves the obstruction in and out of the line, and resist actuation loads.

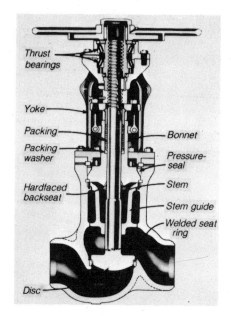

Fig 1: Common valve components

Thrust bearings

Yoke

Packing

Packing washer

Hardfaced backseat

Bonnet

Pressure-seal

Stem

Stem guide

Welded seat ring

Disc

On first thought, a sphere or cylinder seems to be the most economical shape for a valve body intended to resist fluid pressure while open, but there are many other considerations. Many valves require a partition across the valve body to carry the seat opening which is the throttling orifice. With the valve closed, loading on the body is not easy to determine. End connections distort loads on a sphere and other shapes. Therefore, valve bodies tend to be complex shapes with cylinders, rings, intersecting cylinders, solid blocks, and so on.

Passages in the body guide the fluid. Careful design ensures minimum pressure drop through the body, if loss is a concern. Sometimes rigidity and strength are more important, so the passageways are tortuous with abrupt changes to the flow that cause pressure losses. Narrowing of the passageways reduces overall size and cost.

In some valves, the body is split into two halves that bolt together to make periodic disassembly for cleaning and repair easier. Many valves have a bolted cover called the bonnet. It complicates manufacturing, increases valve size, and is a potential leakage path, but sometimes the tradeoffs are worth it.

Valve trim incorporates the fluid-control elements inside. The disc, plug, or ball, depending on the type of valve, is the engineered obstruction. It closes against the seat when the valve is closed. The obstruction is moved by a spindle or stem. Sleeves (cages) are used in certain cases to guide the disc or plug. Of most importance to a valve's performance is the way the disc interfaces with the seat and the relation of disc position to the seat.

Two basic motions are possible, alone or in combination. Either a disc-ported ball or ported plug rotates closely past the seat to produce a change in flow obstruction, or the disc can lift perpendicularly away from the seat so that an annular orifice appears. Ball, plug, and gate valves are of the first variety; globe, check, and safety valves are of the second. A symmetric butterfly valve is an example of a combination. The disc pivots in the seat ring and the opening quickly extends nearly all the way around the seat opening.

How the moveable element and seat interact is vital to valve performance, especially actuation force, the mechanical muscle needed to move the trim, and valve tightness. Wear due to erosion or cavitation may destroy disc and seat surfaces trim because of abnormal pressure drop or the presence of solids in the fluid. Particles which tear at the surfaces embed in the seat/closure area. And when a disc or ball slides past the seat many times under high pressure drop, surfaces can be damaged.

Depending on the valve, the disc connection to the stem may be rigid or loose. In some cases, the seat rings themselves position the disc by an interference fit. In eccentric butterfly valves, gate valves, and plug valves, accurate location of disc, plug, or ball in the stem-axis direction may be important as well. Some motion of the disc/stem connection may be allowed in some designs to help position the disc in the seat. In other cases, too much flexibility will destroy the connection as it flutters under pressure.

The majority of valves with external actuation have penetrations through the body wall or through the bonnet. These may become paths for leakage. Seals and packings are used to block these necessary clearances. Choice depends on valve type, size, and quality. They may range all the way from simple flat gaskets for the body/bonnet area to seal welds and highly engineered pressure seals.

Service conditions and material of construction are closely related, with temperature being the most important factor for steam/condensate systems. Pressure effects are not as much of a burden because they necessitate only changes in wall thicknesses. Allowable pressure decreases with increasing temperature, an effect especially pronounced for nonmetallic materials. Graphitization of low-alloy steels at temperatures above about 775F is a commonly acknowledged limitation (**Fig 2**).

Properties of the construction materials and characteristics of the fluid influence the degree of galling between metal surfaces that contact each other. This is most important in the disc/seat interface.

How flow through a valve changes as the disc approaches the seat is another fundamental element of valve selection. For example, the system's designer needs to know by how much valve flow changes during opening or closing from a partly opened setting. Curves called valve-flow characteristic provide this information.

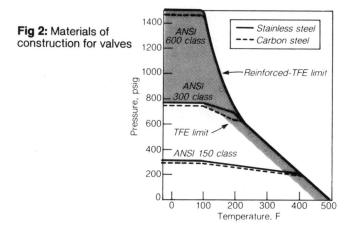

Fig 2: Materials of construction for valves

Repairability and maintainability are factors in valve selection, more so today than ever before. Removing valves for repair in the shop is costly so in-line repair is attractive. Leakage around the disc/seat interface could call for replacement of the seat ring and disc, or for grinding or lapping of the parts.

Chapter ten:
Major categories
of valves

Gate valves

One of the most popular valves for steam condensate systems—or for any plant system, for that matter—is the gate valve. For example, one source claims that 70-80% of the valves purchased for the hydrocarbon-processing industry are gate valves.

A gate valve is designed for on/off service only, or completely opened or completely closed. As such it is used to isolate piping or to block the flow of a fluid as it travels from one piece of equipment to another. Because of its singular service function, the tightness of the seal at the disc/seat interface is the most important design parameter (Fig 3).

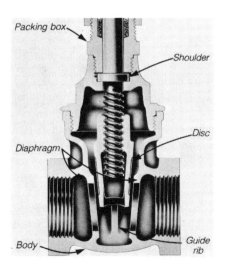

Fig 3: Gate valve

Most gate valves are characterized by the disc and/or by the stem and disc-lifting motion. A disc usually has two faces, both of which engage the seats at closure. The angle of the wedge between the faces helps determine how well the disc seats. Guide ribs are necessary to ensure alignment and prevent it from rotating with the stem.

Gate valves come as either rising or non-rising stems designs. The latter helps keep dirt and other foreign matter from getting into the packing. Also, the non-rising design protects the threads. The two-faced disc, or the wedge, can be made solid or flexible. Here, the design tradeoff is strength vs. flexibility. Flexible wedges are intended to seat more uniformly and rightly under adverse conditions. Depending on how the wedge is cut, though, it may lose some strength (Fig 4, 5).

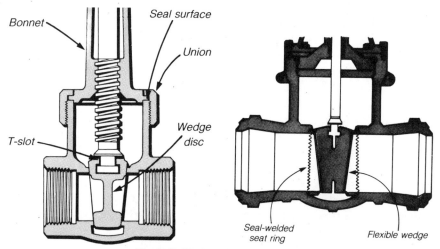

Fig 4, 5: Wedges for gate valves

In larger gate valves, wedges may come as two pieces so each half is free to adapt to its seat ring. Guides on the disc side must engulf the disc guide ribs to prevent excessive spreading apart of the separate pieces.

For steam lines, large gate valves can jam upon cooling if they are closed in the fully heated condition. As the heated body expands, the axial distance between the seats increases. Cooling, combined with the thrust to make it leak-proof, results in enough contraction between seats to prevent opening without applying heat. Though this is more of an operator error, designs are available that compensate for it.

An example is the parallel disc. In one configuration, the wedge surfaces between parallel-faced disc halves are pressed together under stem thrust and then spread the discs apart to seal against the seats. Another approach is to use the upstream pressure to hold the downstream disc against the seat **(Fig 6)**.

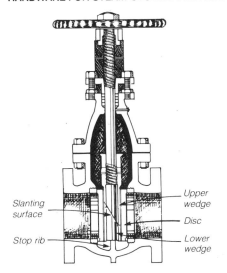

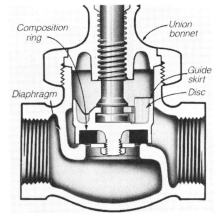

Fig 6: Parallel-disc gate valve   **Fig 7:** Globe valve

Globe valves

For steam-system engineers who are not valve specialists, the globe valve is usually what comes to mind when the subject of valves comes up. Indeed, in one form or the other, it is available for virtually all steam/condensate-system services because it can handle the highest temperatures and pressures involved, it can seal tightly, and it can throttle (**Fig 7**).

Top among the drawbacks of the globe valve is the high friction loss resulting from a flow path that involves two or more turns. Obstructions and discontinuities in the flow path can add to the overall loss. Another short-coming is the large body opening required for installation of the disc. Mounting of the disc on its stem can also pose problems.

Basically, the globe valve operates by the perpendicular movement of the disc away from the seat and orifice. Components must be designed for long-term throttling and high pressure drop.

For steam service to 300 psig, a typical globe valve of bronze construction features a T-shaped body with a Z-shaped bridge wall—a partition across the globular body—that contains the orifice with its edges modified to act as the seat. The stem passes through the bonnet attached to a large opening at the valve top. The bonnet often is of a symmetrical form that eases manufacturing, installation, and repair.

One proven way to ensure tight closure for stop valves in steam and hot-water service is to insert a hard, nonmetallic ring on the disc. Such a ring is resistant to erosion and can cut through solid matter that might have deposited in the seat. Certain changes to the geometry of the seat/disc contact will also promote tight shutoff.

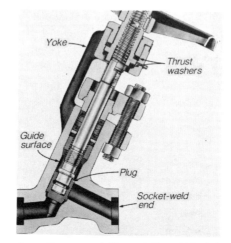

Fig 8: Y-shaped
globe valve

Ways to improve throttling capability while minimizing pressure losses
and effects from pulsation and impact are to lengthen the throttling path or
to consider the Y-shaped body style (Fig 8).

Y-shaped globe valves serve in severe, generally high-pressure service. In
small sizes for intermittent flows, pressure loss may not be as important as
other considerations that favor the Y-form. Y-shaped globe valves feature a
seat and stem angled at about 45 deg, the optimum compromise between
straightness of flow at full opening and need to accommodate the stem,
bonnet, and packing in a pressure-resistant envelope.

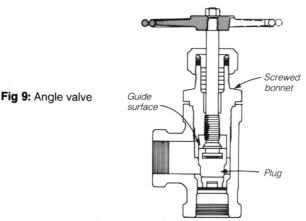

Fig 9: Angle valve

Closely related to the Y-shaped and T-shaped globe valve is the angle valve
(Fig 9). With the ends at right angles, the bridge wall of an angle valve can be
a simple flat plate. Fluid flows through a single 90-deg turn, discharging
downward more symmetrically than the discharge from an ordinary globe
valve. Installation advantages alone may favor the angle valve; it can, for

example, eliminate an elbow in a piping system. Discharge characteristics from angle valves are so favorable from fluid-dynamics and erosion standpoints that many control valves have this configuration (control valves are discussed as a separate subject later on).

Regardless of body form, globe and angle valves have several aspects in common. Disc travel in a globe valve is nearly always short compared to other types. Theoretical lift height needed to form the circular-aperture area equal in area to the seat orifice is one-quarter of the seat-orifice diameter, and provides ample discharge area for throttling.

In the basic configuration, the disc is a cantilever mounted at the stem end. As such, it is sensitive to fluid turbulence, especially if the stem is long and slender. Heavy stems cause difficulties with the packing box. Slight flexibility in the stem is usually advisable.

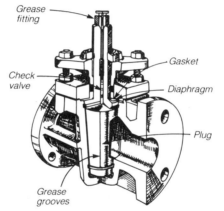

Fig 10: Plug valve

Plug valves

Some in the valve business feel that the plug valve is possibly the oldest control valve, tracing it back to the water systems built during the early Roman Empire. Common plug valves feature a transverse hole in a cylinder or truncated cone rotating in a closely fitting body cavity that aligns with the body inlet and outlet passages when the valve is opened. When turned a quarter turn, or 90 deg for shutoff, the plug presents a solid-surface obstruction to the fluid. Differences in design and detail for the available plug valves are built on this simple concept (Fig 10).

Advantages of the plug valve are: It has low flow resistance, if the size of the port is near the line size, it can throttle on moderately demanding services, its straight passages prevent accumulation of sediment or scale, it needs a minimum of installation space, it exhibits fast response, and it helps break down high pressure drops because of the two symmetrical orifices.

Clearances and leak prevention are major considerations with plug valves. Because many of them are of all-metal construction, lubrication is needed in many cases to keep actuation torque at reasonable levels and prevent galling. Regular maintenance is a must with the lubricated version. In steam/condensate systems these valves can be used for water only if lubricant contamination is not a serious danger. For the most part, however, these valves are used to handle gases, fuels, and other hydrocarbons.

Use of materials such as TFE and nylon extends the range of the plug valve for steam/condensate service as long as the temperature remains under the safe limit for the material. If so, then these materials help solve the sealing problem for the metal plug without lubrication(Figure 11).

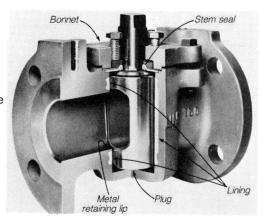

Fig 11: Plug valve with lining

An important characteristic of plug valves is their adaptability to multiport construction, or three- and four-way valves. A plug can be designed to replace as many as four conventional shut off valves. Extra body outlets and L-shaped plug passages are available as well.

Ball valves

There are two main classes of ball valves—the floating-ball type and the trunnion type, often with metal seat rings. Floating-ball-valve operation is similar to plug-valve operation. A through-ported ball, supported by and rotating in two TFE seat rings, one around each body outlet, allows flow when the ball port aligns with the seat rings. Upon a quarter turn, the spherical surface of the ball seals against one or both rings under the fluid pressure on the ball.

Although the spherical is most common by far (Fig12), a conical surface near the ball is possible, too. It reduces the pressure on the seat and prevents seat cutting by the leading edge of the ball port as the valve closes.

Torque for actuation is transmitted by a stem with a rectangular head that fits closely in a slot on the ball top. Slot alignment is always at right angles to the passage in the ball for slight movement and tight sealing.

Changes and improvements in the floating-ball valve include the seat

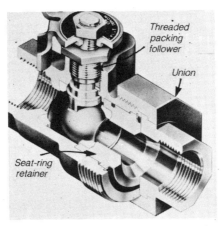

Fig 12: Spherical ball valve

rings. Usually, the ring is nearly triangular in cross-section, though there are departures. Sealing against fluid pressure may be at both rings or only on the downstream ring. In the latter case, the rings include grooves on the outer rim to let fluid pass the upstream ring after slightly pushing the upstream face of the ring away from the ring seat in the body.

Pressure ratings for ball valves handling room-temperature water go to 4500 psig for sizes to 3/4 inches, and to 300 psig for sizes up to 2 inches with special nylon seats.

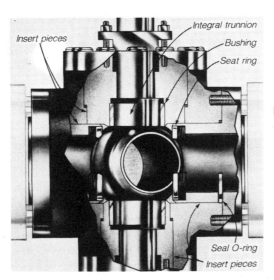

Fig 13: Ball valve set in trunnions

For more support and positioning of the ball in larger sizes, the ball must turn on trunnions (Fig 13). Trunnion stubs may be external, protruding into the ball, or integral with the ball.

These types of ball valves go up to 48 inches in line size. They are often used for high-pressure gas-transmission lines or low-pressure water lines. The ball is usually extensively hollowed out in these applications. As long as seat-ring pressure is controlled, torque for breakaway and running tends to be lower for the trunnion type than for the floating-ball type. Rolling contact bearings on trunnions have reduced the torque even further.

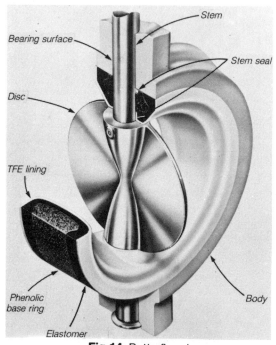

Fig 14: Butterfly valve

Butterfly valves

Another very old and versatile type of valve is the butterfly valve (Fig 14). Application range of the butterfly has been extended in recent years due to design modifications and materials improvements that allow it to shut tightly.

Flow is controlled with a circular disc that pivots, in a horizontal or vertical configuration, perpendicularly to the direction of flow within the pipe. In the fully opened position, the disc is parallel with the longitudinal axis of the pipe. Closed, the disc blocks the flow. When between the two positions, throttling capability is achieved.

Among its most prominent advantages are low maintenance, because of the relatively few moving parts, and savings in space and weight. These advantages stand out even more in larger sizes. What the butterfly valve was not capable of for many years was tight shutoff.

Older designs feature a roughly symmetrical disc journalled in bearings that turn in or just outside of a short ring-shaped body. When constructed of all-metal components, tight closure is nearly impossible because of the abrading drag of the disc edge past the body seat ring. Usually, another valve must be in line for tight shutoff. New seating materials for the symmetrical butterfly valve have advanced its closure capability. An elastomer seat, for example, allows it to close tightly on low temperature/pressure, up to about 180F and 200 psig, which is adequate for the most demanding powerplant cooling-water service. Excessive water velocity in symmetrical designs can be a problem because it can lift up the seat ring of the nearly closed valve and wreck it. Maintenance usually involves taking the valve out of the line and dismantling it completely, even just to replace an O-ring. Often, the valve is discarded to avoid repair failure.

Another class of butterfly is the eccentric or offset design. It was developed to improve upon the symmetrical type. Here, the disc turns on a shaft or stem which is displaced from the disc/seat contact surface. This results in a continuous sealing area, unbroken by a stem penetration that can cause discontinuities in line-sealing action and present problems in stem sealing. Also, the seat ring itself can be very narrow, and can be shielded against fluid effects; this widens the choice of soft seat materials and the manageable pressures and temperature ranges.

Flow through all eccentric butterfly valves is disturbed more than in the symmetrical type (Fig 15). Net fluid pressure on the disc tends to close a partly opened valve, so actuation mechanism must be able to hold the valve at a setting. Also, the eccentric type is not fully bidirectional. Pressure ratings for

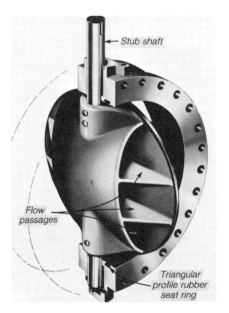

Fig 15: Symmetric butterfly valve

leak-proof service and flow resistance differ depending on the direction. Performance tends to peak at 3750 psig or 1500F.

Development of the eccentric butterfly follows two principal lines. In the first, rubber seats seal against rounded edges or slightly conical tapering disc edges. This variety is suited to water service at pressures up to 200 psig and in the largest sizes needed in a steam/condensate system. In the larger sizes, the disc must resist heavy water-pressure load, but not be bulky enough to impede flow significantly when open.

The other line of development is disc-edge sealing surface on a sphere centered close to where the shaft crosses the piping axis. One way to picture this variety is to imagine a ball valve with only a very thin piece of the ball remaining around one port. Opening motion of the disc draws the edge away from the seat ring. Sometimes, there's a second offset of the shaft axis to reduce dragging at the start of opening.

Pressure classes for the offset butterfly are more or less closely fitted to ANSI 150 and 300, with about 720 psig the limit for tight shutoff and reasonable seat-ring life in water and ambient-to-warm temperatures. Some can go to ANSI 600 in pressure. On higher temperature liquids, pressure ratings drop until the limit of between 450 and 500F is reached. For higher temperatures, metal seatings may be needed but this approach has not yet been successful.

Diaphragm and pinch valves

Diaphragm and pinch valves are not normally used for steam/condensate handling. They are applied to auxiliary subsystems of the steam-generating unit such as the water-treatment system.

Instead of using controlled geometry and rigid materials for closure, the diaphragm valve has a highly flexible and extendable elastomer sheet forced down into a seat. Its chief application is in water-treatment work for handling chemicals, corrosive liquids, and liquids that might be contaminated by metal contact. In these services, its body is made of a rigid plastic or is metal-lined with a corrosion-resistant material. Diaphragm valves can also serve as relief valves, especially in corrosive service.

The cushioning effect of the elastomer also tends to protect the valve from solids in the fluid, as it seals tightly. This makes it suitable for slurry and waste water applications. At the same time, the weakness of the design is the elastomer diaphragm, which can crack from repeated flexing or be torn by sharp-edged solids.

Extending the concept of the diaphragm valve leads to the pinch valve, also used for slurries, ash disposal, fuel-slurry handling, and water and wastewater treatment. Here, closure occurs through squeezing together elastomer walls over a wide area, rather than against a narrow metal contact ridge.

Check valves

Another class of valve dating back hundreds of years is the check valve. It differs from the others already discussed in that it is not actuated outside of the pressure envelope, and the actuation is not within the control of the operator. Rather, opening and closing is governed by the flow direction. The singular duty of a check valve is to prevent the reversal of flow in a pipeline.

A disc, ball or plate unseats in response to the differential pressure across the valve opening. Pressure of the fluid provides some of the actuating force; the rest comes either from gravity, a spring, or other means.

Check valves must open easily and completely to pass flow with little pressure drop or disturbance to flow patterns. They must resist damage to seat disc and disc hinging or guiding means in all flow conditions, and close quickly, without injuring the valve, at the instant of flow stoppage.

Some leakage, such as in discharge lines from pumps, often is permissible in a check valve. On the other hand, it should not leak if it is intended to isolate a machine for maintenance, and there is not another isolation valve in line.

Reliability has a higher meaning with check valves. In addition to closing, they must close according to the desired position/time relationship. Many check valves have the characteristic of lag in closing so that flow actually stops before the plate closes completely. If it closes suddenly and late, widespread damage can occur to the valve and the related piping and supports due to water hammer.

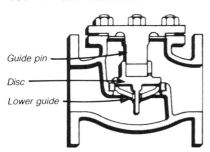
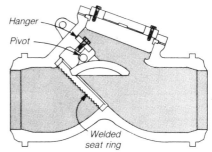

Fig 16: Lift-type check valve   **Fig 17:** Tilting-disc check valve

Flow through a check is a source of energy loss. Cost evaluation of the valve should take this into account. If the check valve has a perpendicular lift from the seat, it is a lift check (Fig 16); a pivoted lift and a swing check are the two basic categories. A third, less common, motion is rotation of an eccentrically mounted butterfly-disc seating (Fig 17) in a slanted seat (tilting disc).

In many cases, the check valve comes as part of a control system. These units are powered by a stoppage of flow, sensed or programmed deliberately. Here, the valve can look more like a control valve than a conventional check valve.

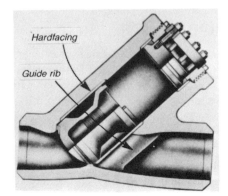

Fig 18: Y-body
check valve

Using a Y-body (Fig 18) with hardface, accurate guide surfaces for the heavy plug is one way to reduce the flow resistance and pressure drop in large lift checks. Internal dashpots are another accessory to prevent damage from water hammer caused by sudden closure after backflow has begun.

One interesting modification of the swing-disc check valve is a dual-disc arrangement where two D-shaped pieces pivot about a vertical pin and seat on flat D-shaped surfaces. They can close with low shock.

Check valves are sometimes adaptable to shutoff or even moderate throttling service. Boiler stop-check valves are an example of shutoff service. Sometimes, throttling and shutoff is accomplished by combining the check valve with another valve—such as a globe or an angle valve—in one package. Boiler nonreturn valves are one application where this is done.

Some general factors in check-valve selection are presented here. 1) Fluid type. Water and other liquids may require swing checks or special types of lift checks. Gases, on the other hand, tend toward lift and tilting-disc designs. 2) Need for tight sealing. The ability to repair or replace a disc or seal without removing the valve from the line is an important consideration. 3) Flow direction. Horizontal flow is the simplest, upward vertical the next simplest. Downward vertical and oblique flow should be closely analyzed. 4) Leakage. If leakage to the outside is of prime importance, a pressure-seal or seal-welded version should be considered over a check valve with bolted bonnet and body penetrations for the shaft. 5) Position. If it is essential to determine position and exercise the valve, then a penetration of some kind into the valve body will be necessary.

Safety and relief valves

As their names imply, safety and relief valves operate automatically and must offer high reliability. A feature common to most safety valves is a spring that holds the disc down on the seat until the pressure inside the protected equipment reaches a preselected value and overcomes the spring force, opening the valve.

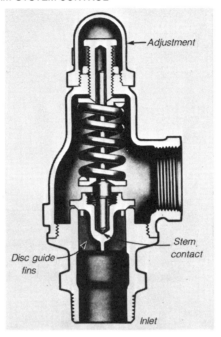

Fig 19: Safety valve

Liquid service is less difficult than gas service. Liquid pressure drops off quickly when a small amount is bled off, so relief and safety valves can be comparatively small (Fig 19). The valve mechanism can be totally enclosed with a removable cap over the adjustment for spring compression. Alloy construction may be necessary for trim, spring, and/or body in high-temperature service.

Relieving pressure in steam is more complicated. In steam/condensate systems, services include boilers, superheaters, piping headers, and accumulators. Relief flow must continue for some time to have an effect on the expanding fluid. Also, steam leaking through a slightly opened seat is damaging, so the valve must have quick and full opening slightly above operating range, and a quick total closure when the pressure falls below the set pressure. To conserve energy, the blowdown (pressure drop between set pressure and shut off pressure) from the relief valve must be minimal; high-pressure valves may have blowdowns of 3-5% of the operating pressure.

Closing of the valve is sometimes assisted by pressure of steam in chambers above the guided disc (Fig 20); the steam can also assist in opening the valve if eductor tubes are used (Fig 21). To prevent leakage from the discharge space, past the stem, or from drain holes, a bellows is sometimes used to seal off the stem passage. This approach is common in nuclear work. Even slight leakage of high-pressure steam can cause velocity erosion of the seating surfaces.

Further difficulties with steam leakage may call for a more complicated design. One approach is to use a separate, small pressure-controlled valve or

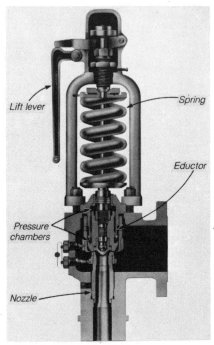

Fig 20: Safety valve with steam-assist closing

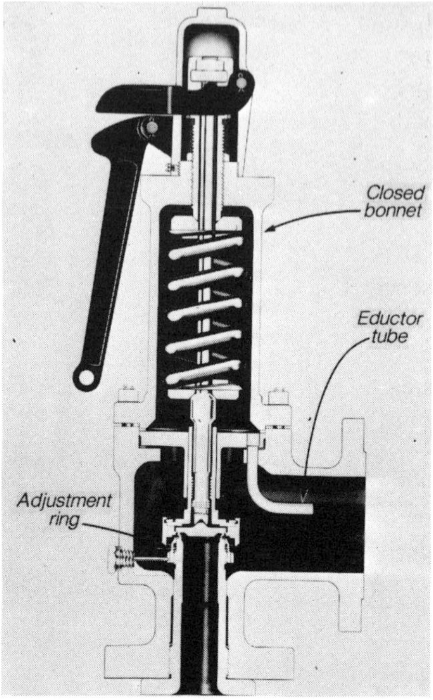

Fig 21: Safety valve with steam-assist opening

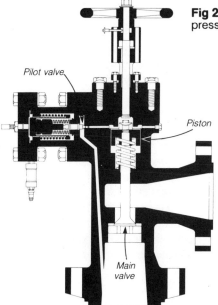

Fig 22: Safety valve with small pressure-control-valve assist

Fig 23: Flow vs control-valve-stem travel

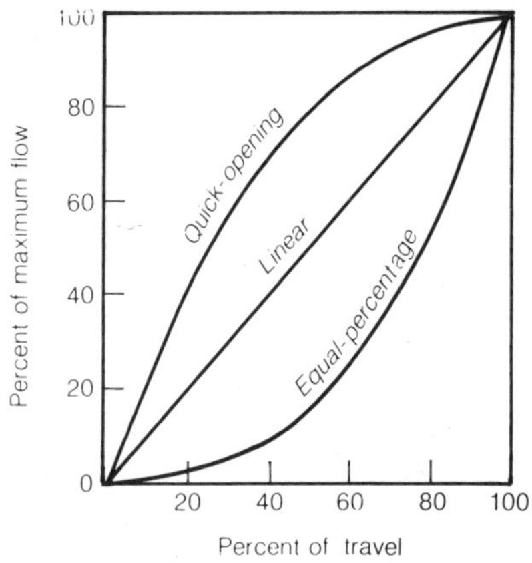

a control system external to the valve to actuate a release of vessel steam pressure from a chamber bounded by a piston on the disc (Fig 22). The vessel pressure, which also acts on the other side of the piston, forces the main valve open.

Instrument valves

All the valve types described up to now can be made in smaller sizes to service the small-bore piping and tubing that links main pipes and components with instruments and control devices. These small valves serve as pilot devices for larger valves, for lubrication of large equipment, for water and steam sampling, for draining, venting, and bleeding, and for continuous blowdown of small boilers.

Selecting and specifying these valves and controllers should not be relegated to the back burner. Failures in instrumentation piping mean switching to inefficient manual control until the problem is fixed. There are many design operating and maintenance factors peculiar to the range of small valves and control systems which should be thoroughly reviewed in any comprehensive valve program.

Control valves

Barring the gate, check, and relief valves discussed earlier, all the other types can serve as control valves. A control valve is power-actuated to throttle or modulate flow reliably for long periods of time without loss of stability. Sometimes, the phrase designates special on-off valves that are power-actuated to meet requirements such as actuating speed and position sequencing. Auxiliary devices that accompany control valves include positioners, transmitters, transducers, and amplifiers. These are discussed in separate sections.

As electronics and microprocessor technology impacts the control systems, there undoubtedly will always be a need for the full-size control valve as a final-control element.

Certain concepts are common to all control valves:
1) Valve-flow coefficient. This indicates capacity and is represented by the flow in gal/min of 60F water through the valve at 1-psi pressure drop, at a given upstream pressure. It is usually given for the fully opened position.
2) Flow characteristic. This is the relationship of how the flow, expressed as a percentage of the maximum, varies with stem travel (Fig 23). It is based on a constant pressure differential across the valve, or on the actual differential pressure during operation. A linear characteristic has advantages for the entire control system, but an equal-percentage control valve will often give an approximately linear system characteristic. Some valves fall between the two. Near the valve's full closing point, performance tends to deviate from

the general form of the flow characteristic, often between 0.5 and 5% of full flow.

3) Rangeability is the ratio of maximum controllable flow to minimum flow. It can vary from 200 to 20 or less, and is important for judging the valve's performance at less-than-design flows, especially very low flows.

4) Pressure recovery is important where cavitation is possible. It is a measure of how much the outlet pressure increases after passing through the high-velocity seat area. High-recovery valves tend to cavitate at lower pressure drops. Usually, recovery is higher for ball and butterfly valves than for globes.

5) Leakage is often stated as a percentage of the full-rated flow. It is usually higher for double-seated valves than for single-seated ones.

6) Unbalance on a plug results from pressure and geometry differences on the upstream and downstream sides of plug or disc. Difficult to eliminate entirely, unbalance requires greater actuation force.

7) Noise and cavitation characteristics are important factors today. Two general methods of countering these effects are breaking up the flow into many smaller streams, and/or forcing the fluid through repeated contractions and turns. The drawbacks of these modifications, more common with larger valves, are added leak paths, plugging, and premature wear of multiple thin-wall components.

8) Quick opening characteristic, common to all globe valves with a flat disc and seat, is not as useful in control work as it is in other services.

Requirement for actuation force is another practical aspect of control-valve selection. If inlet flow tends to push the valve disc open, the actuator must exert heavy force to hold the valve closed. If, on the other hand, the flow is down through the seat and fluid pressure holds the disc shut, actuation to initiate opening is correspondingly higher. There are various ways to balance fluid forces on the disc.

Friction in stem packing, journals, and seats of ball, butterfly, and plug valves, and at the piston rings of actuator cylinders is another concern in control-valve design. High friction may dictate use of a positioner, a device that senses the percentage of valve opening and signals the actuator to bring the opening to the new level called for by the control system (to be discussed in detail later).

Whether a control valve is balanced or unbalanced is another important design aspect. In the unbalanced design, usually a single-seated top-guided globe valve (Fig 24), the fluid pressure acts on the exposed underside plug area. In a balanced design, such as a double-seated plug with top and bottom guides (Fig 25), fluid pressure acts on both sides of the plug and the forces tend to cancel. While a balanced valve can operate with a smaller actuator for a given size, the valve usually exhibits a higher leakage rate. Another difference is that fluid flows upward through an unbalanced valve and usually flows downward through the balanced trim. Balanced designs have

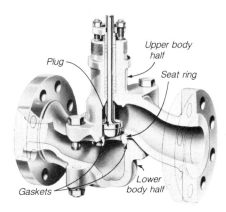

Fig 24: Unbalanced control valve, above and top right

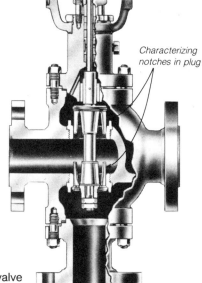

Fig 25: Balanced control valve

shutoff pressures ranging from 10-50 times as high as those of unbalanced designs using the same actuator, while maintaining the same leakage rate. All unbalanced valves—whether globe, butterfly, ball, or plug—tend toward dynamic instability, sometimes called chatter, under high pressure drops as the valve approaches the closed position.

Split-body globe valves offer ease of disassembly for maintenance, inspection, cleaning, or repair. A reduced-trim feature—a seat ring that is replaceable with one of reduced-orifice size—helps eliminate instability and wear caused by throttling near the seat.

Cages, as part of a control valve, not only help hold the seat rings firmly and guide the plugs, but, more importantly, help to characterize flow, if the valve is equipped with several tapered ports. If flow is through a cage and downward past the plug, the annular space around the cage must be large enough to distribute flow evenly. Holes must be numerous enough to bring fluid uniformly into the cage. There are many special-purpose variations in cage holes.

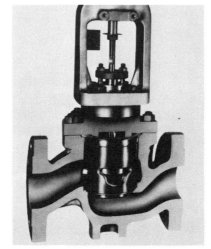

Fig 26: Cage-type control valve with balancing holes in plug

Balancing the plug in a cage-type valve can be accomplished with holes in the plug (Fig 26). But this makes it necessary to prevent leakage past plug outside diameter when the valve is closed. Cage valves with pilots in the plug or with seal rings are possible options to reduce the chances for leakage. In the pilot-valve design, as the stem lifts, the pilot valve opens first and vents the space above the plug, which has become pressurized by leakage past the clearances or rings.

An overriding advantage of the cage design is the ability to alter the valve

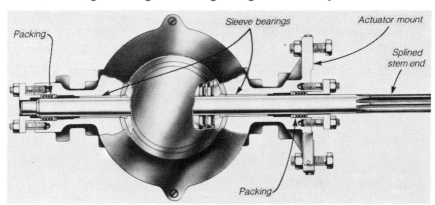

Fig 27: Butterfly control valve

performance by changing the cage and trim within the same body.

Another common control valve is the butterfly. Symmetric butterfly valves with metal-seating surfaces may have low rangeability because of clearances; a shutoff valve is therefore necessary to close the line completely (Fig 27). Soft-seat butterfly valves can also have a control function but they are known to experience seat damage during long-term, low-flow throttling—at least in some designs. Eccentric butterfly valves with a control function are more common. Often exactly like the on/off versions, they can also be equipped with heavier stems, longer stem bearings, or special packing boxes when used as control valves. Rangeability is as high as 100:1. Ball, diaphragm, and pinch valves can also be control valves, but they are not prevalent in steam/condensate systems. In addition, there are a host of unusual control-valve designs that are very different from the ones described here. They are often called upon for very special services.

Regulators

Regulators are small automatic-control systems that can be used to control the flow of relatively clean fluids. Reducing regulators control downstream pressure, temperature, or flow rate while the upstream pressure varies widely; backpressure regulators control upstream pressure. Often, the measure of control achieved by a regulator is not as precise as that achieved by the more extensive system of a control valve, actuator, and controller. But the controlled variable can be held remarkably constant over a full flow range.

The key to sizing and selecting a pressure regulator is the proportional band, also called the percent of deviation, or droop. This is the change in pressure necessary to effect regulator operation. To illustrate: If a pressure-reducing regulator is to maintain downstream pressure at 100 psig, a 5% droop implies that 95-100 psig is the process tolerance that will be achieved.

Self-contained pressure regulators are operated by the resultant of forces created by the displacement of internal springs and the products of pressure and the areas of the diaphragms on which the pressure acts. They eliminate separate power supplies, external signal-conditioning devices, most external piping,and possible damage caused by a loss of compressed air or electric power. Other advantages are close-to-instantaneous response, simpler maintenance because of the fewer moving parts, and less chance for leakage since most have no external packing glands.

The simplest pressure regulator is the direct-operated version, sometimes called the one-stage version (Fig 28). It senses pressure on one side of the diaphragm and has an opposing spring or constant air pressure on the other. Set point is adjusted by changing the compression force on the spring or the loading pressure. Greater spring forces raise downstream pressure. A more complicated type, known as the two-stage version, uses a pilot mechanism to

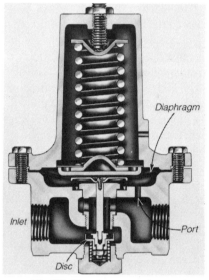

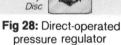

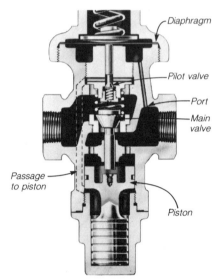

Fig 28: Direct-operated
pressure regulator

Fig 29: Pilot-assisted
pressure regulator

sense changes in set point, then correspondingly load or unload a diaphragm
or piston. The piston has enough area to operate the large main valve
(Fig 29).

Diaphragm motion of the single-stage regulator is restricted, therefore,
capacities are limited. For example, in steam service, 1000 lb/hr is close to
the maximum for many available units.

Error or droop of the main valve is not as pronounced in the two-stage
regulator. Droops for single-stage are 15-25%, for double-stage more like
1-5%. Two-stage pilot-operated models can handle steam flows of up to
several hundred thousand pounds an hour.

For regulating temperature, a two-stage, pilot-operated type of regulator
comes equipped with a temperature-sensing element that produces enough
force to drive a pilot valve's actuator diaphragm. Popular sensors for this
duty are thermocouples, bimetallic sensors comprised of dual strips with
different coefficients of expansion under heat, resistance-temperature detectors, and thermistors.

Digital valves

Digital valves have been available for at least 15 years, but for a variety of
reasons, they have not captured any services typical of the steam/condensate
system. In this rather novel approach to fluid control, the valve body contains a series of differently sized on/off orifices in binary sequence——
1,2,4,8,16, and so on. Each orifice requires its own plug, seat, and actuator

coupled to an air or electronic signal. Signals do not have to be modulated or closely controlled.

What the digital valve offers in a control system is precise flow changes and rapid response—over the full range setting without overshoot—if this is needed. High rangeability is possible with a large number of elements and it does not change from valve to valve because it is not dependent on the seat and disc configuration. Hysteresis and actuator friction have little effect on valve performance either.

One general disadvantage of control systems containing digital valves is that even minor signal errors and out-of-phase difficulties can cause mistaken settings that deliver far more fluid than desired. With conventional valves, minor signal errors only cause minor errors in plug position. Disadvantages for steam systems are twofold. For one thing, the initial cost has traditionally been higher than that of the valves discussed earlier; this is because the valve involves a complicated form and a multiplicity of components. For another, it does not serve well for high-temperature steam. There are usually temperature limits of the tight-sealing seats that preclude its economic use with the steam conditions found in most utility and industrial applications. But digital valves have found successful application in natural-gas-handling systems.

Chapter eleven: Principal valve components and auxiliary hardware

Actuators

When a valve is broken down into its components, the actuator heads the list in importance. Actuators convert energy into the work required to move the valve stem which opens, closes, or holds the valve at a throttling position. Although the balance of control systems are undergoing miniaturization, the actuator is not likely to follow suit. Positioning of a valve against friction and fluid forces will always require heavy thrust or torque.

Just about all valves require an actuator of some sort. Even many check and safety valves need one so they can be tested. Trends toward plant automation, reliability, and reduction of cost and of the physical effort of plant operators constantly expand the demands on the design of the actuator. Even the traditional simple manual actuators can be made complicated by the addition of insulation, enhanced grips, or ability to impart a hammerblow. Larger sizes, harsher environments, high actuating forces with their negative side effects, and requirements for fail-safe operating modes have complicated the design of the many actuators that move the valve stem automatically.

Details of the valve design have as much to do with matching an actuator to the valve as with the choice of manual or automatic operation and the choice of air, electric power, or oil as the energy source. Fluid-power-type actuators are the most important for steam/condensate systems. Air is more common than oil. Sometimes the line fluid can supply motive power to either diaphragm or piston actuators, but this is more common for water and gas distribution systems than for steam systems.

Basic types of actuators include manual, diaphragm, piston, cylinder, rotary, vane, electric-motor, solonoid, and electrohydraulic.

MANUAL ACTUATORS

The simplest types of manual actuator are handwheels and levers. A hammerblow feature can be included to open large valves. Symmetric lugs on a sleeve can move freely through about a quarter circle, building up inertia before they strike matching lugs on the valve yoke bushing. Lever handles are standard on small quarter-turn valves and quick-opening gate valves. Low-priced ball and butterfly valves may have cast stop lugs on the body to engage lugs on a pressed-steel lever. Lever length is usually not more than 12 in. for valves up to six inches in size.

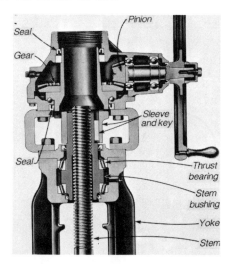

Fig 30: Actuator with gearbox

To muliply valve-stem torque or thrust so operators can more easily open large valves, a simple gearbox can be set on the yoke top of a gate or globe valve or on the bonnet of a ball or butterfly valve (Fig 30). Sealing of gearboxes is important to protect bearings and gears.

Threaded spindles and various types of linkages are another way to rotate a quarter-turn valve. Linkages not only convert linear thrust to rotary motion but also help in varying torque or thrust over the valve-stem travel range.

Several types of valves—wedge-gate, plug, ball, butterfly, and taper-plug—require more breakaway torque than mid-stroke torque. Seat/disc interface and shutoff pressure can vary, too, making actuator requirements hard to specify for both opening and closure seating. Increased seating load is desirable with globe valves, which can leak if not tightly closed. Also, direction of fluid pressure, whether it acts against the globe-valve disc or in the same direction, influences the amount of closing thrust needed.

For large actuators in high-pressure valve service, the input torque and number of turns needed to stroke a valve completely may be so high that an auxiliary power source is needed. Remote manual actuation may also be necessary. The chain wheel and the stem extension with its universal joints are the simplest examples.

DIAPHRAGM ACTUATORS

The diaphragm actuator, usually with a return spring, but occasionally without, is a simple and widespread form (Fig 31). It can deliver enough thrust and stroke for the majority of steam/condensate system applications. Nearly all these actuators are air-powered, although operating with hydraulic oil is not excluded, just rare.

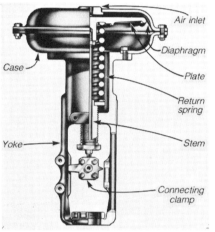

Fig 31: Diaphragm actuator

An important advantage of the diaphragm actuator is the absence of friction of piston rings. Sealing of the air chamber is by a flexible diaphragm with most of its flat area supported by a plate. The unsupported annulus between the plate and case can resist air pressure up to about 60 psig.

A helical spring can be added to a diaphragm actuator to allow it to develop a return force proportional to the travel of the diaphragm plate and stem. In theory, increasing air pressure will move a valve stem linearly, allowing the control-valve characteristic to modify flow as desired. Side effects in the actuator should be considered, however.

The valve is called normally closed if the return spring pushes the valve shut.

If valve-stem-to-packing friction is low, and fluid effects aren't too variable, a diaphragm and spring actuator will take a position which is nearly directly proportional to air pressure. This is advantageous in control-system design.

One limiting factor of diaphragm actuators is a high initial spring-compression force when high closing force is necessary, such as with a globe valve with flow upward against the disc. This means higher air pressure to move the diaphragm plate and stem.

Many times a manual handwheel provides emergency override actuation of a diaphragm-actuated valve.

For heavy thrusts, diaphragm actuators of exceptionally heavy construction are available. One approach is the tandem configuration.

Diaphragm actuators are normally limited to strokes of about three inches or less. This is enough for most globe and cage valves, but could be too short for rotary-motion valves, where high torques require long lever arms and therefore long actuator strokes.

Hydraulic snubbers can be considered if instability in the valve is a problem. The snubber can consist of a hydraulic cylinder added between the diaphragm case and spring housing.

PISTON (CYLINDER) ACTUATORS

Compact and able to deliver long strokes and heavy thrusts, the piston actuator (Fig 32) plays an important part in fluid handling. It is most com-

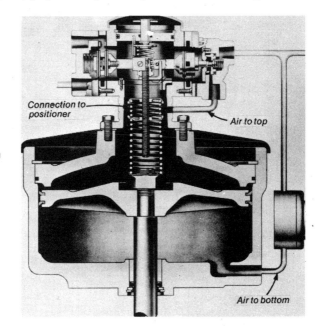

Fig 32: Piston actuator

mon on ball and butterfly valves. Some piston actuators simplify piping problems in new and existing systems being equipped with power-actuated valves.

The fact that pneumatic piston actuators accept plant air pressures often simplifies valving and control problems in on/off applications. Hydraulic versions are usually smaller for a given thrust. Both can handle the highest steam-system design-pressure differentials adequately. They are, however, more expensive than diaphragm actuators.

Linkage or gearing often is necessary for quarter-turn-valve actuators. Gearing (rack and pinion) is far more common for small piston actuators. Simplest variety is the conventional double-acting cylinder, pivoted at one end and driving directly to a crank on the valve stem. The pivot pin attaches to an arm mounted on the valve topworks, or to a sleeve bolted to the line pipe. Many variations on the basic concept are possible.

Cam action converts straight piston motion to a quarter turn in several actuators. In particular, this type of action has been applied to butterfly valves and small valves. Yoke and paddle mechanisms are also used to move the valve. One feature of the paddle mechanism is that it gives an initial opening torque higher than the final closing torque.

When gearing is used in cylinder actuators, it is usually of the rack and pinion variety. Return springs are also common, but not with large hydraulic actuators because the springs become too large. How the spring is attached to the piston differentiates many offerings.

Cylinders for quarter-turn valves are sometimes tandem or parallel units with two cylinders or two pistons acting together. Enhanced thrust is partially compensated for by the increased height, which affects mounting and installation.

Mounting of piston actuators can be a problem, especially if yokes and frames are heavy, and/or where torque is heavy. Valve-bonnet design is also affected. For example, top-entry ball valves offer a broad support base, while the flange at the stem top in butterfly valves may not support long horizontal-cylinder actuators.

Because most cylinders in these actuators depend on piston seals to prevent escape of motive fluid, friction of the ring can vary. It can also be aggravated by dirt and sludge in the air or oil lines.

ROTARY-VANE ACTUATORS

Torque produced by fluid pressure on a vane or arm coupled to a valve stem can actuate any type of quarter-turn valve in many sizes. Linkage and gearing are eliminated, but torque can only be varied by changing the fluid pressure. This is the concept behind the rotary-vane actuator.

This actuator is another air-powered type. A paddle-like vane in a sector-shaped casing is pushed through a quarter-turn arc. Two vanes in a semicircular case can double the torque (Fig 33). To increase the operating angle, there is a rotary-vane-actuator assembly with two housings. Spring-return action in a rotary-vane actuator is possible, usually through a clock-spring type of arrangement.

ELECTRIC-MOTOR ACTUATORS

Actuators described previously have comparatively short strokes or turn

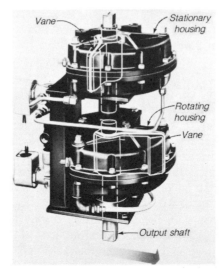

Fig 33: Rotary-vane actuator

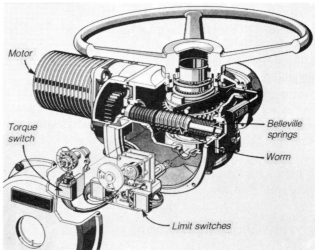

Fig 34: Electric-motor actuator

angles. If continuous rotation is desirable, as in gate and globe valves with screwed-thread-actuated stems, then the actuator (Fig 34) must be either an electric motor (used in almost all cases) or an air-powered motor.

Electric motors are inherently high-speed, low-torque devices, so speed reduction is usually furnished by the actuator. Assembly and mounting of the motor and speed-reduction unit differs widely.

Electric-motor actuators can cost considerably more than common air-powered units; this is, of course, very decisive in final selection. For large or high-pressure valves, where a pneumatic or hydraulic actuator might be a special design or require its own compressor or pump system, the electric motor tends to be competitive.

Many small electric-motor actuators are designed for automated or computer-controlled systems and work as on/off or modulating devices. They are intended to make the control system all-electric, with actuator and control integrated to varying extents.

While some electric-motor actuators are suitable for hazardous locations, many industries do not use them in this service because of a lack of confidence in electrical systems external to the actuator.

Although most electric-motor actuators are for intermittent service, they can be used in modulating control. Desirable motor characteristics in modulating service include low-rotor inertia, low maximum-current draw, and rapid rise to speed when the valve is partly opened. In modulating control, frequency of actuation can be as high as 60 times per minute.

More conventionally, higher rotor inertia is needed to deliver the heavy blows required for gate- or globe-valve seats. Rotor inertia is magnified by the gear reduction. This indicates a need for travel and torque limit. Most of these types of actuators have travel-limit switches to stop motion and perhaps apply a brake when the valve stem reaches a selected point.

Torque limitation is considerably more involved. Cutting off the motor current at preselected increases in sensed torque may not be enough. Brakes and spring compensators may also be necessary, expanding the actuator package and increasing its price. Fortunately, the valve-damage problem can usually be controlled by standard torque-limiting electric-motor actuators. Nuclear-power steam systems require the highest technology of torque limitation. In general, torque limitation is more of a requirement for protection of motor-actuated gate and globe valves than for quarter-turn valves.

Other motor characteristics relevant to actuator service are the pattern of torque variation over the stroke (important in gate and globe valves), and the possible need for extra control provisions for motors modulating control—otherwise hunting, a cause of motor overheating, may occur.

SOLENOID ACTUATORS

The solenoid valve, confined to globe-type valves, is the simplest electric-actuated valve (Fig 35). Though stroke and thrust are very limited, speed is fast, often a fraction of a second between opening and closing. Pilot-operated and direct-operated models are available. The former opens or closes a small valve in the main globe-valve disc to release pressure from the disc top, enabling stream pressure to open a large main disc.

In both, the solenoid, usually located in the valve body or bonnet, pulls directly on the valve disc. Pilot-operated solenoid actuators handle larger flows, but neither type is capable of throttling flow, acting only in the fully opened or fully closed position.

ELECTROHYDRAULIC ACTUATORS

The electrohydraulic (EH) actuator is more expensive than the actuators discussed above. It is used chiefly in applications that the conventional

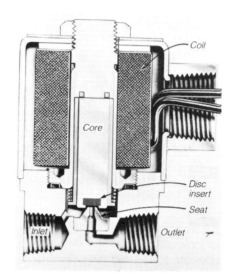

Fig 35: Solenoid actuator

actuator can't handle, where a combination of difficult specifications needs to be met. If high thrust or long stroke are the only requirements, hydraulic or electric-motor actuators can do the job. Similarly, if a fast stroke is needed, pneumatic actuators are available.

But if a valve with high thrust and a long stroke delivered in a few seconds is required, the EH actuator is the best/only option. The EH actuator consists of a hydraulic-cylinder actuator, a nearby oil supply, and adequate controls to develop the desired performance in thrust, speed, stiffness, frequency response, and other characteristics. Most of these EH actuators have the oil-pressurizing system packaged with the actuator, enlarging package dimensions. Such a configuration differs from a simple hydraulic actuator in performance. Electrohydraulic units serve both on/off and modulating applications. The cost disadvantage for on/off uses is normally greater than for the modulating services.

Oil supply for a EH actuator may come from centrifugal gear, vane, or axial-piston pumps. In most systems, the pump is fixed-delivery, with oil flow controlled by external valving or by on/off operation.

The cylinder itself is almost a double-acting type, even for designs in which a return spring strokes the valve to close or open setting. Vertical cylinders are most common. In several systems the rod extends out both cylinder ends to allow a direct mechanical link with the position transducer.

Position and speed of the modulating piston are controlled by the servo system, which includes a position transducer or positioner that detects piston position. In some systems a linear voltage-differential transducer supplies a control signal to a servo valve; others rely on a combination of nozzles, flappers, and torque motors.

Pressurized reservoirs and accumulators are needed in some systems.

These serve three functions: maintaining pressure on the oil system to prevent cavitation or air entry, developing temporary high flow rate from a small hydraulic pump, and actuating the valve should electric power to the pump motor fail. One drawback of accumulators is that they can occupy a considerable fraction of the total space.

Performance of EH actuators varies widely, depending on fluid-system design objectives. Small actuators of standardized design used in large-scale production can deliver up to 16,000-lb thrust. Most applications for this class, however, are well below 10,000 lb. Large custom-designed actuators for specific applications can deliver over 300,000-lb thrust in on/off mode.

Although stroke length for the standardized designs can extend to 24 in., most models give no more than 4 in. This is adequate for globe and quarter-turn valves to about 16-in. size. Stroke length for large special actuators is practically unlimited, and easily sufficient for the largest gate valves.

Speed of the EH actuator is a tradeoff against thrust and, to some extent, stroke. The standardized designs may have valves ranging from 0.1 to 2 in./sec, while valves as fast as 12 in./sec are available from special designs.

All EH actuators are very stiff against varying fluid forces inside the valve, and actuator motion is smooth. Dynamic performance such as frequency response can be tailored to meet system needs.

Valve/actuator interface

The valve's requirement's for thrust or torque and for stroke length or angle dictate choice of actuator and specification of performance. Pressure of the fluid, differential pressure across the valve seat, area of seat, dynamic effects of flowing fluid, and friction are the main factors influencing thrust and torque; valve size and type obviously influence stroke length or angle.

Since friction and dynamic effects are hard to calculate, the specifier must usually rely on data from the valve supplier. Seemingly minor design differences or manufacturing tolerances can influence thrust or torque needs, while maximum cost benefit depends on knowing how valve type and design affects actuator sizing.

One familiar on/off control element is the gate valve. Its actuator has to overcome not only the frictional force on the seat, caused by upstream pressure or wedging, but also friction in the screw threads of the stem, also a significant and more constant load over the stroke. The yoke bushing, transferring thrust reaction to the valve structure must rotate in either rolling-contact bearings or plain bearings. Lubrication is advisable in either case. Packing friction adds slightly to the thrust load, and the line pressure acting on the stem area produces a force opposing seating and assisting opening.

In most gate valves, the disc seating surfaces slide for a short distance along the seat as the valve begins to open. Sliding should stop when the disc wings contact the sides of vertical recesses in the body. Further upward motion will result in friction along the recess slide-ways. The sliding force decreases as Dp (differential pressure) across the disc drops.

A high wedging force, which will add to thrust at the next opening, can result from closing a gate valve at high speed. To overcome this excess wedging action, many gate-valve actuators include the ability to deliver an initial hammerblow at start of opening.

Compared with a gate valve of the same seat-port diameter, a globe valve and the similar angle valve have low stem travel. Lift for on/off valves is usually only a quarter of seat-port diameter. Wedging action sometimes can occur here as well. Direction of flow makes a large difference in required thrust.

In both gate and globe valves with screw-thread actuation of the stem, thread friction, plus wedging action in the seat, is often sufficient to prevent opening of the closed valve. A globe valve actuated by diaphragm or piston actuator, and with upstream pressure tending to open the valve, requires either a spring or pressure in the cylinder to keep the valve closed under pressure.

The ball valve differs from gate and globe valves from the actuation standpoint. In the floating-ball type, common in sizes to about 6 in., actuation load comes chiefly from friction of the ball against seat rings of Teflon, fiber-glass reinforced Teflon, or Buna N. Teflon gives lowest friction coefficients, reinforced Teflon is higher, and Buna N. is highest.

Seat friction and torque are influenced by abrasives in the fluid and fluid temperature. A valve left open or closed for long periods in fluids that can cause encrustation on the ball will sometimes cause an increase in breakaway friction.

Stem effects are another source of actuation torque. Friction can result from the tendency of internal fluid pressure to force the stem outward against a restraining thrust washer. Substituting a Teflon washer for an asbestos one will decrease friction, though not eliminate it entirely. Stem-seal friction or stuffing-box friction also contributes to torque.

The bearings used to position the ball in trunnion-type ball valves can be a source of friction. As springs and fluid pressure force the seat rings against the ball surface, bearing friction occurs. There may also be stem-seal friction and often stem-thrust friction.

In symmetrical butterfly valves, torque is required to overcome interference fits between the disc and the flexible liner, friction at stem-seals, stem thrust, and fluid dynamic forces. Seat interference is adjustable in large butterfly valves. With metal-to-metal-seat valves, some leakage is usually permissible, so stem effects and dynamic effects are a bigger fraction of the whole.

In eccentric or offset butterfly valves, by contrast, stem effects are comparatively small. This is because the seat ring contacts the disc uniformly and fluid pressure from either side gives added sealing force. Stem effects are comparatively small. Breakaway torque values for these valves are in general higher than for symmetrical butterflies but lower than for ball valves. There is considerable variation from design to design, however, and safety factors also vary.

Since fluid flow tends to move all butterfly-valve discs, especially eccentric types, the actuator must be able to hold the disc in the selected position.

Plug valves can have straight or taper plugs. Both kinds may have lubricated metal-to-metal contact, with the plug lifted slightly--this reduces torque during operation. In most taper-plug valves, adjusting screws or similar devices allow the downward thrust to be changed.

In control valves, modifications have three purposes: reduction of actuator force, lessening of wear in the throttle setting, or improvement of resistance to dynamic forces.

One way to control plug movement better without increasing actuator force is to enclose the globe-valve plug in a cylindrical cage. In balanced-cage valves, however, sealing-ring friction cancels some of the saving in actuator force. Small stems and spring-loaded Teflon rings will reduce packing friction in many globe-type control valves.

Quarter-turn control valves include ball and butterfly. In several types of ball and plug valves, the moving closure element is cut away so that high seat-ring friction occurs only near shutoff.

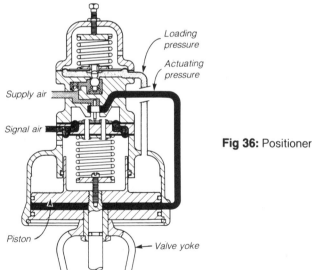

Fig 36: Positioner

Positioners

As mentioned previously, some actuators have no return spring. These actuators need some device that will sense present valve position, compare it with the desired valve position, and transmit a signal or control air pressure to move the actuator. Positioners, which can be either pneumatic or electro-pneumatic, perform this function (Fig 36).

A simple positioner works like this: A change in signal pressure changes the setting of a flexibly mounted bellows capsule or diaphragm. The capsule then moves a small valve that admits more or less supply air to one side of the piston. A feedback spring between the piston and bellows capsule restores the setting of the small valve. This gives the positioner a linear characteristic. In some cases, a non-linear can be used instead to characterize the flow vs signal relationship. Cams offer flexibility because they can be cut to give the valve a different characteristic; this avoids any changes to the trim.

Positioners can be either force- or motion-balanced. Force-balanced designs develop a feedback force with valve position to balance the controller signal acting on an input diaphragm or bellows. Motion-balanced ones design function by comparing valve-stem movement with a similar motion caused by bellows expansion from an air signal.

The electropneumatic positoner has a similar function, but it works quite differently. It converts an electrical input signal to an electromotive force that must be balanced by deflection of a spring when the valve end moves.

Mounting of the positioner is an important factor in selection. Positioners integral with actuator are compact and better suited for short-stroke valves. Externally-mounted positioners have the advantage of handling long strokes more easily; they are also simpler to open when adjusting control-system performance parameters such as span.

One major disadvantage of positoners is that they sometimes degrade dynamic performance of the control loop if the positioner/actuator frequency response is too close to that of the overall control system. Under these circumstances, volume booster might be preferable, if speed is the primary concern.

Packing & Seals

Most types of fluid-handling equipment, including valves, require an opening in the pressure body for the movement of internal closure and control elements.

One such opening is the bonnet. But the bonnet and other lesser openings as well need a physical barrier between the stationary wall and the moving shaft. Otherwise, inevitable clearances between the opening and the stem or spindle allow intolerable leakage. This is where packing and seals come into play.

Valve size and pressure are important factors when choosing among the wide variety of bonnet closures. For small valves, a gasket that seals against the clearances of ordinary threads of a simple screwed-on bonnet is usually sufficient, though a union bonnet is sometimes a better alternative. Both, however, require a round closure which is awkward for larger valves. Bolted

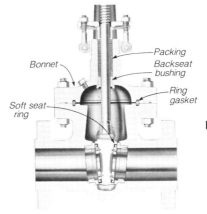

Fig 37: Bolted bonnet

bonnets are adequate for large valves under moderate pressures (Fig 37), and small high-pressure valves. Pressure is the critical factor in gasket selection. Two popular choices are flexible metal/asbestos and metal profile rings.

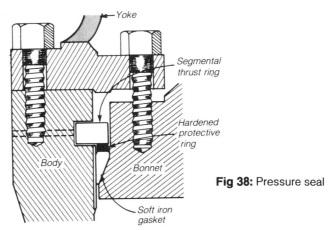

Fig 38: Pressure seal

Several types of pressure seal (Fig 38) are available for ultra high-pressure applications. Here, internal pressure wedges two sloping surfaces together. Damage-free assembly of pressure seals—especially in large bonnets—requires that the bonnet's upward thrust be resisted uniformly, that jamming be prevented, and that actuation reactions be transferred into the valve body or the piping. If properly designed and installed, they can handle the highest pressure of a steam system. One condition to watch for is bonnet overpressurization. This occurs when a gate valve seals on both faces of the disc and traps liquid in the bonnet. Upon slight heating, the liquid expands, placing undue load on the gasket and stem packing. Schemes are available to prevent this.

Sealing of the stem against fluid escape is a troublesome problem, depending on the stem movement of the valve. Long-sliding motion, such as with gate valves, or helical motion, such as with globe valves, is harder on the stem-seal system than the quarter turn of the small ball valve.

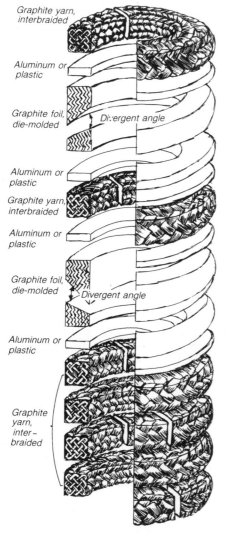

Graphite yarn, interbraided

Aluminum or plastic

Graphite foil, die-molded

Divergent angle

Aluminum or plastic

Graphite yarn, interbraided

Aluminum or plastic

Graphite foil, die-molded

Divergent angle

Aluminum or plastic

Graphite yarn, inter-braided

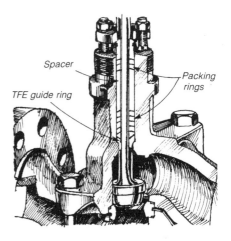

Spacer

TFE guide ring

Packing rings

Fig 39: Compression packing, left; Automatic packing rings, above

Two types of packing are available: jam (or compression) packing and automatic packing (Fig 39). The first is a material that is confined and squeezed into interference with the rotating or sliding-shaft stem. Automatic packing is a special, yielding profile ring with interference fit on a stem.

Packing is available in a wide range of materials and forms. These include asbestos, carbon, graphite, tetra-fluorocarbon (TFE), Aramid, glass fiber, and metals alone or in combination. Choice must be based on pressure, temperature, corrosion nature of the fluid or environment, resistance to abrasion, resistance to chemical attack, and ability to dissipate heat.

Whatever material is chosen, it must be largely impermeable to the fluid that it is sealing. Packing made up of fibers, yarns, roving, or other forms contains many passages through the packing between neighboring fibers and yarns. Although compression of the packing will eliminate most of these, enough will remain in connected form to cause excessive leakage unless special materials are used in the packing to decrease its permeability.

Special methods of intertwining the component strands of packing are used to increase resistance to physical damage such as abrasion and cutting. Braiding is the most common method. Before and after braiding the yarns themselves may be impregnated with oil and graphite particles. This reduces damage during braiding and lubricates the fibers and yarns in service. So can abrasion resistance, which can be achieved either by use of an inherently abrasion-resistant material, or by addition during braiding of abrasion-resistant yarns that end up along the edges.

Regulation of compression on valve packing can be a problem, too, especially if the aim is to be in the narrow range between heavy stem friction and leakage.

This requires periodic adjustment of the gland setting, which can result in considerable expense and time if the number of valves is large. One solution is an automatic device to maintain constant gland pressure. Such a device must take into account the heavy force needed and the small amount of motion to produce the needed increase in force.

Tolerances in the box and associated components are another consideration affecting packing performance. One important dimension is the clearance between stem and valve bonnet at the bottom of the stuffing box. Some clearance is necessary here, but any more than a few thousandths of an inch carries with it the danger of extruding certain types of packing down into the valve if gland pressure is excessive.

One way to prevent this without expensive close tolerance of the valve-bonnet aperture is the bottom ring, sometimes called a junk ring. In addition, the device prevents transfer of abrasive solids up into the packing on valves whose stems slide or move helically upward.

Another added component in a stuffing box is a lantern ring (Fig 40). It is a metal ring grooved and bored to allow flow of cooling liquid or lubricant around the shaft or stem, and can replace several rings of packing in a deep stuffing box.

Lantern rings have another function. If water enters at a pressure above that in the valve the higher pressure will keep the internal fluid out of the packing and prevent harm to the packing from chemicals or abrasives. In this function, the ring is usually near the box bottom.

Valve packing can take the form of rings cut from a length, rings molded

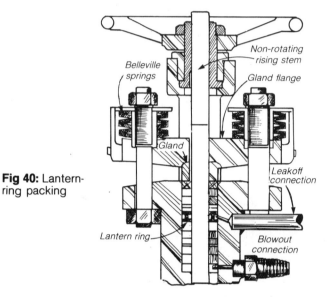

Fig 40: Lantern-ring packing

Non-rotating rising stem

Belleville springs

Gland flange

Gland

Leakoff connection

Lantern ring

Blowout connection

at the factory but split, or continuous formed rings. The last-mentioned variety, common now for small valves, requires valve handwheel removal, but this is not a serious drawback with a small valve.

Packing requirements vary depending on the type of valve. Because of the small size and low number of packing rings in a soft-seat ball valve, more expensive graphite foil is frequently selected. For globe- or cage-type control valves, the friction of the packing can be more troublesome than in manually operated valves. Stuffing boxes are often deep for more flexibility in accommodating various systems of packing and sealing.

Since large-scale dismantling of valves is undesirable, packing should be capable of installation with only the gland and bolts removed. The removal of old packing is often difficult, particularly if an asbestos ring has wedged into the bottom of a deep packing box. Nevertheless, it is always necessary to make sure that the last trace of old packing is out; otherwise, the new packing may fail in a short time.

AUTOMATIC PACKING

One of the simplest and most effective examples of automatic packing for limited speeds and motions is the O-ring (Fig 41). Obviously, this elastomeric device is fragile and calls for close tolerances for the ring itself and for the groove into which it fits. The pressure inside forces the ring against one side of the groove, deforming it to block the clearance. A lubricated O-ring can slide along a surface and can rotate against it, but heat buildup and wear place limits on the allowable surface speed.

Some of the drawbacks and limitations of the O-ring are common to all automatic packing for fluid-handling equipment. For example, elastomers

Fig 41: O-ring automatic packing

can be easily cut by nicks or burrs on shafts. Pump shafts and valve stems tend to have long lengths across which the O-ring must be pushed to reach its operating groove. Considerable care is demanded when installing an O-ring in the field, especially when it is a replacement.

Automatic packings are successful in actuator cylinders and shafts. There, the tolerances are easily met, lateral movement under load is low, sliding speeds are moderate, and lubrication is generally available. Many small valve actuators have O-ring piston seals, and O-ring piston-rod seals are also common on large and small cylinder actuators.

As pressure increases and tolerances open up, it becomes increasingly necessary to prevent extrusion of an O-ring. A myriad of devices, ranging from a simple metal backup ring to composite rings of two materials, are available.

V-shaped rings of several configurations are another answer where automatic action is wanted (Fig 42). Pressure of either the fluid or a spring makes the initial interference fit even tighter during operation. The shape of

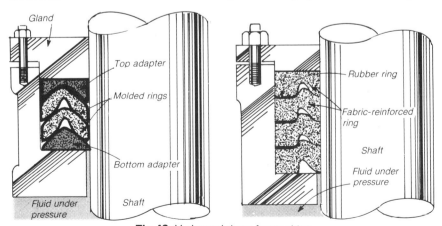

Fig 42: V-shaped rings for packing

the ring determines whether the frictional contact will be much higher when the ring is under internal pressure. The degree of interference fit is influential, too.

A seal that depends on initial compression into a confined groove or other space to seal is called a compression seal. High friction and wear are characteristic, even though many of this type, such as O-rings, operate for years in what appear to be difficult services. Another class of seal, the lip seal, has two wings that can be forced outward against the shaft or enclosing sleeve by pressurized fluid. The initial interference fit can compress the lips to a varying degree to give zero-pressure sealing, but the lip seal is basically not as effective as the compression seal at no pressure.

For other applications, however, the lip seal has proved very successful in cylinder actuators, where tolerances can be made and maintained closely. The lip seal is effective in preventing escape of air or oil. Just enough pressure of the lip on the moving part to prevent more than a tiny amount of oil being trapped under the contact surface assures lubrication, low friction, and low wear.

Because of the light contact of lip seals, they are sometimes added to the outside of jam packing or mechanical seals to prevent entry of dirt or liquids, or to act as a seal to retain flushing liquids. A lip seal will also protect an actuator cylinder from outside dirt by acting as a wiper on the piston rod.

Automatic packing for valves has had considerable success in plug, ball, and butterfly valves, in which the stem is closely restrained and rotates only 90 or 180 deg. O-rings are often the only seal against escape of fluid. With such materials as Viton, temperature can go up to 500F, and resistance is assured to all chemicals that the valves' stem and internals can resist.

In most steam/condensate valves, however, automatic packings are at a disadvantage because of deposit buildup, lack of lubrication during operation, and lateral movement of the shafts and stems. Successful application depends on taking these factors fully into account.

Controllers

Controllers are used to couple a controlled process variable to the manipulated variable, usually governed by the control valves (Fig 43). In some instances, the controller input is mechanically linked to a controlled variable indicator powered only by the output of the measuring element. Such a design is called a self-operated controller. Sometimes, the sensing element does not develop enough power to move the control valve, so a power amplifier is used.

Controllers can be self-operated, pneumatic, electronic, or hydraulic. Most types are either on/off (also called two-position) controllers, or proportional controllers, but specialty units may also include rate control and reset capability.

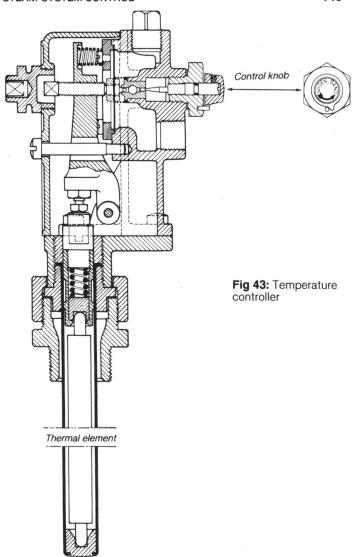

Fig 43: Temperature controller

Control knob

Thermal element

Electronic controllers are used more widely today than pneumatic ones because they are more easily linked to process-control computers, replace lengthy pneumatic transmission lines with electrical wiring, and eliminate pneumatic time lags. In addition, they afford more accurate readings and have fewer moving parts. Their chief disadvantage is that they need special equipment designs to meet safety and hazardous-environment requirements.

The heart of the electronic controller is a high-gain operational amplifier which accepts dc signals at the input and supplies dc signal at the output. If the amplifier itself uses ac signals, they must be converted back to dc signals. Most will accept a variety of input signals by properly sizing input resistors, but the output is usually limited to a single range. Transducers are used in the field installation to convert the actual physical measurement, such as pressure, to an electric signal, and then to convert the electronic output signal to a pneumatic one powering the control valve or final control element.

Pneumatic controllers supply controlled pressurized air--usually in the 3-to-15-psig range--to a pneumatic valve actuator in response to an error signal based on the deviation of the measured variable from the set point. A variety of pneumatic-equipment configurations are available to produce the controller output corresponding to on/off, proportional, reset, and rate-control modes.

In on/off controllers, control action is outputted to the valves either at zero or at a maximum value. This mode is commonly used in temperature-control systems, such as when fuel to the furnace is either flowing or not flowing, or when the current in an electric-heating coil is either at the maximum or not delivered at all. It is also used in pressure-relief valves and thermocouple-operated safety valves.

The action of the proportional controller is smooth and continuous over the operating range. Under steady-state conditions, the controller assumes an intermediate position where the measured variable equals the set point except for offset. Output varies in an amount proportional, either directly or indirectly, to the deviation of the measured process variable from the set point. A reset feature and/or a rate control can be added to a proportional controller; with these additions, it is commonly called a two-mode or three-mode controller.

Hydraulic controllers are not as common as the other three. They are similar to pneumatic controllers, but have some important advantages and disadvantages of their own. Advantages include a high ratio of output power to input power and faster response times since the fluids are essentially incompressible. The main disadvantage is the increased complexity of the hydraulic fluid-handling and control system. Sometimes fluid pressure aids or supplements the sealing fit.

Steam traps

Until energy management became prevalent, failed steam traps in a large industrial complex could, in the extreme case, waste the output of one of several steam boilers. Since the 1970s, however, the steam trap, which helps conserve steam, has had its prominence elevated. As a result, today there exists a separate technology built up around its design, operation, and maintenance.

The steam trap has three main functions: Allow condensate resulting from steam giving up its latent heat to flow to a collection system; vent air and other gases to maintain steam temperature and reduce corrosion; and prevent escape of steam, forcing it to lose all its latent heat. Traps prevent steam loss that sometimes can be as high as loss from an open safety valve.

Though steam-trap technology has made great strides in recent years, proper design is not enough. Operating conditions are equally important for ensuring proper performance, reliability, and long life. Obviously, this is true to some extent for all equipment but perhaps is even more so for traps. The sheer number of these devices in a powerplant or industrial complex—a large refinery may have over 20,000—makes them hard to manage, and there is a tendency to overlook effects from water treatment, startup procedures, process operation, and weather when specifying, purchasing, installing, and maintaining traps.

To perform its function, a trap must be able to distinguish steam from condensate and non-condensible gases, open to discharge the condensate, then automatically reclose before any live steam is lost. Steam traps can be divided into three main types based on the methods used to detect the difference between steam and condensate. Float traps sense density difference; thermostatic traps sense temperature drop between condensate out of contact with the steam and the steam itself; thermodynamic traps sense differences in velocity, and therefore static pressure change between flow of water and steam through a restriction. A fourth category, fixed orifices, also differentiates by allowing much more water to discharge than steam.

FLOAT TRAPS

Float-trap design is based on the ability of a float to detect easily the level of condensate in a vessel. A common float trap (Fig 44) consists of a small chamber containing a float, the outlet valve, and some form of linkage. When condensate drops into the chamber, the linkage multiplies the float's buoyancy force enough to open the valve against the pressure drop across the valve.

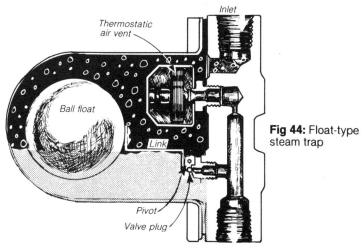

Inlet

Thermostatic air vent

Ball float

Link

Fig 44: Float-type steam trap

Pivot

Valve plug

The inlet of the float trap is almost always near the top. The outlet is necessarily near the bottom but not at the lowest point. The piping connections are often horizontal and on the body side, rather than the vertical inline connections.

The condensate enters at the top, and cascades down the trap-body wall and into the bottom pool. If the pool is at a low level and the trap is lightly loaded, solids and sludge will have a chance to drop out to the body bottom.

Increase in condensate level gradually lifts the float, which is a light sheet-metal hollow ball. Simple leverage enables the float to overcome pressure difference between the steam system and the discharge line, and slowly open the trap valve enough for a discharge rate equal to incoming condensate flow.

A sudden load increase on the steam system will increase condensate flow and open the valve further, while decreased load will begin to close it. At all times, however, the valve must be below the water surface to prevent direct loss of steam.

Two facts should be immediately clear from this. First, the float trap must be installed in only one attitude--with its valve down. Second, the trap modulates discharge instead of opening and closing intermittently.

One advantage of a conventional float trap is that it can start totally dry--unprimed. When dry, the float is down, closing the valve. Even if the valve leaks somewhat, the trap body will slowly fill until the water level covers the valve and it begins normal operation.

There are, of course, some drawbacks. If solids and sludge build up in the trap-body bottom, they can prevent the float from sinking and closing the valve. This is apt to happen in systems with dirty steam, poorly done piping repairs, or corrosion in steam-using equipment. A screen or strainer upstream of the trap will stop much of the solids from entering; those that do enter can be blown out through the small plugged outlet at the bottom of many float traps.

Another problem is the entry of air and other noncondensible gases into a trap. In the simple float trap, in which the valve is located below water level, gas cannot leave except in solution. Because little gas is in solution with condensate at the desired few degrees below saturation, a separate means of venting gas is necessary.

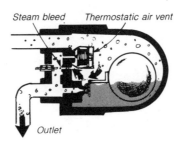

Fig 45: Float and thermostatic steam trap

Steam bleed   Thermostatic air vent

Outlet

The thermostatic air vent is so frequently used for this that the trap is usually called "float and thermostatic" (Fig 45). The vent, which consists of a small valve and an actuator that responds to temperature, is normally located high on the trap body and close to the inlet.

If the air vent opens too close to the temperature of saturated steam, trap loss will be excessive. Because gas in the steam lowers the steam's partial pressure and temperature by a few degrees in actual service, a small amount of gas tends to be constantly in the steam space of the trap. To vent all the gas would call for a thermostatic-vent setting so high that live steam would blow to waste if no gas were present.

Another means of releasing gas is to use a small manually operated valve, either alone or as an aid to a thermostatic vent.

Ordinary float and thermostatic traps have capacities to about 7000 lb/hr at 100-psig inlet pressure. Larger models have 10 times this capacity, with a pressure limit around 250 psig. As the trap size increases, pressure limit decreases because float buoyancy plus the single stage of leverage cannot overcome a high differential pressure on a large valve orifice. Compound linkage can increase the pressure limit, but it is more complicated and expensive.

Comparatively large body size is another drawback of the float and thermostatic trap. But its outstanding weakness is vulnerability to water hammer. Pressure shock entering the trap in a slug of condensate can crush the thin-walled float, causing the trap to fail closed and flood up to the thermostatic-vent level. The thermostatic vent will then provide condensate removal at reduced temperature and capacity.

INVERTED-BUCKET TRAPS

The idea of allowing a bucket inside a trap body to fill with condensate and sink the trap, pulling on a linkage connected to the trap valve, is an old one. Originally the bucket was open at the top, but an inverted cylindrical bucket in the body has now replaced this practically obsolete design. The inverted-bucket trap (Fig 46) is basically a cyclic one, alternately opening and closing abruptly. Its cycle begins with some condensate in the bottom of the body and enough hot steam in the bucket to lift the bucket in the condensate and close the trap. Continuing gravity flow of hot condensate into the bucket through the inlet tube would pressurize the steam in the bucket, assuming that none of the steam in the bucket condensed, and finally stop flow, causing condensate to back up in the inlet pipe.

Cooling of the trap body and the condensate in it would eventually condense the steam in the bucket. A much faster way is to let the trapped steam escape through a vent hole in the inverted bucket. The steam, with entrained air, rises through the water in the trap to the body top, where it may condense or remain depending on condensate temperature.

When enough steam has escaped to raise the level of condensate inside the

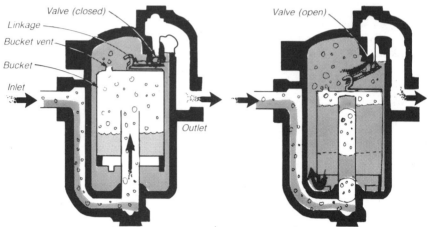

Fig 46: Inverted-bucket steam trap

bucket to the point where the bucket, which is weighted at the bottom, is heavier than the water it displaces, the bucket sinks and tugs down on the linkage connecting it to the valve plug at the trap top.

The valve will not open, however, until the condensate level in the bucket is high enough to enable the bucket's weight in the condensate to pull the valve open. Pressure differential, valve orifice area, bucket weight, and mechanical advantage of the linkage are all factors here.

Trap opening is sudden and complete, discharge is rapid. Condensate is driven by the steam-system pressure into the inlet tube, down the bucket interior sides, and up the annulus between bucket and body. Water and steam at the body top are discharged in a turbulent scrubbing flow. The pressure reduction along the flow path can cause some condensate to flash into steam.

When enough steam has collected again in the bucket, the bucket becomes buoyant, rises, and quickly closes the valve, completing the cycle.

Linkage can be fixed-ratio or a two-stage design with higher leverage at the start. Occasionally, when higher force is needed for higher steam pressure or large valve orifice, the linkage is compound.

Capacities of inverted-bucket traps top out at 18,000 lb/hr at 100-psig inlet pressure for iron-construction traps. Traps of steel or special construction can handle higher flows and pressures to 2,500 psig.

THERMOSTATIC TRAPS

This type of trap includes several different operating mechanisms, all of which are in service today in various applications. The thermostatic trap opens and closes by means of a force developed by a temperature-sensitive actuator (Fig 47). The controlling temperature can be a constant one, or one which varies depending on conditions in the steam system. One basic problem for all thermostatic traps is that of keeping actuating temperature close

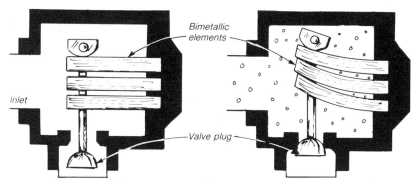

Fig 47: temperature-sensitive actuator for steam trap

to the saturation temperature so that condensate will be hot but no live steam will blow out of the trap.

The saturation temperature of water increases as pressure increases, rapidly at first, then more slowly. Condensate discharged slightly below saturation-line temperature assures that the steam has yielded its latent heat and is still hot enough to prevent solution of O_2 or CO_2.

Discharge of condensate considerably under the saturation-line temperature at a given steam-system pressure creates a problem. Although the condensate has given up more heat, there is also the possibility that O_2 and CO_2 have dissolved in the water, making it corrosive. Also, the condensate may back up to an impermissible level in upstream equipment, reducing heat transfer.

For these reasons, most thermostatic-trap designs aim at keeping trap-opening temperature close to the saturation line. A safety factor may be necessary, because a trap opening at a temperature above the saturation curve is blowing live steam along with the condensate.

THERMODYNAMIC TRAPS

This class of trap has earned its name by taking advantage of the relatively higher velocity created by passage of hot condensate or steam through a narrow gap, as contrasted with the lower velocity for cooler water. The two chief varieties are the disc trap and the impulse trap. In a less common variety, the lever trap, the action of a small disc trap inside the trap body does the piloting for the opening of the trap's main valve.

The disc trap (Fig 48), is a small compact trap with only one moving part—the flat disc that rests on the valve seat and is both the valve plug and the actuating piston.

In startup of a steam system, the initial cold condensate can push up the disc and spill into an annular channel around the inlet passage. From the annular channel, one or more outlet passages lead down to the trap outlet.

The valve seat ring, which separates the inlet from the annular channel, is narrow so that pressure drop across it is low. When condensate has warmed

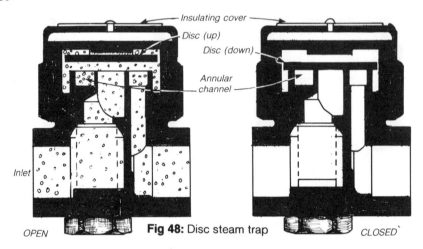

Fig 48: Disc steam trap

OPEN   CLOSED`

up sufficiently, the pressure drop is enough to cause the hot water to flash into steam some of which collects on top of the disc. The steam flow on the underside of the disc is at higher velocity than the water flow, and, therefore, its static pressure falls more, in accordance with the Bernoulli equation.

In the rapid and positive motion, these two effects close the disc with a snap action. The upstream steam pressure, acting on the small inlet area on the disc, is not enough to counterbalance the steam pressure above the disc. The trap therefore stays closed until the steam above the disc condenses and lets condensate push open the disc again.

One deficiency of the disc trap is poor gas handling. Large amounts of air will close the trap, because the velocity, when air and CO_2 are flowing in it, resembles that of steam. This means that another air-removal means is necessary for effective startup of a steam system being drained by disc traps. The analogy with the float trap is obvious, although the air-relieving means is not usually in the trap body. In the impulse trap (Fig 49), a constant small

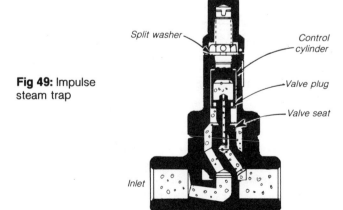

Fig 49: Impulse steam trap

flow of condensate (or steam, if load is very low) slips past a narrow piston skirt on the valve plug and escapes out through a narrow passage in the plug. Cold condensate creates unbalance pressure on the piston undersurface and lifts the valve plug. Hot condensate flashes and the steam pressurizes the piston top and closes the valve.

FIXED-ORIFICE TRAPS

The fixed orifice (Fig 50) and its several modifications have long been part of condensate systems. Its operation is based on the principle that the flow rate (lb/hr) through an orifice is much higher for liquids than for steam. As a result, a properly sized orifice can drain a constant condensate load without loss of steam.

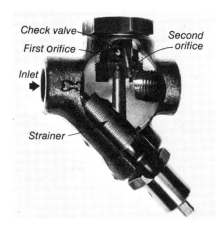

Fig 50: Fixed-orifice trap

Application of the fixed orifice to condensate drainage has provoked controversy. This centers on steam loss through an orifice imperfectly matched to system condensate load or draining a system with widely varying load. Because the fixed orifice is necessarily smaller than the intermittently opened valve orifice of a steam trap, the sensitivity of the fixed orifice to blockage by solids or consolidating sludges is another hotly discussed question.

OPERATING CONDITIONS

The life of steam traps can be prolonged by attention to effects from plant conditions other than condensate load. Five effects that deserve further investigation are corrosion, flow effects, thermal effects, solid and liquid contaminants, and mechanical stresses.

Corrosion in traps results chiefly from the action of O_2 and CO_2 dissolved in the condensate, though other gases present in chemical- and metallurgical-plant steam systems can also cause problems. Even good water treatment can't prevent O_2 and CO_2 from entering during the early stages of a shutdown, for example. At this time, pools of water collect at low points and corrode the surroundings. Preventing corrosion is a prime factor in trap selection.

Flow effects can also damage traps. Hot condensate flow through a trap may flash and cavitate in the valve and in internal passages. Metal erosion will result from sputtering in the flow, especially with droplets in the fluid, and from high-velocity steam flowing in a failed open trap. Shaking loose linkage components is another problem.

Heat and cold, inside and surrounding traps, is another source of potential damage. Thermal cycling occurs during system startup and shutdown, and during normal operation, condensate temperature changes by as much as a hundred degrees. Traps also differ in resistance to freezing weather. Freezing occurs in the traps as well. These and other thermal effects impair trap life.

Solid contaminant problems can be partially relieved by using screens and straining. Upstream piping sections with heavy solid deposits are particularly dangerous; if sludge dislodges and pours, it will clog strainers to uselessness.

Finally, vibration of internal linkage and valve stems through the connected piping affects the trap. Secure mounting can correct some ill effects.

Steam-metering devices

Measuring steam flow in steam/condensate systems involves two conflicting demands. On one hand, for most closed-loop control systems, good repeatability is more essential than absolute accuracy, since the control system tends to compensate for variations. On the other, energy-management applications demand meters with absolute accuracy.

The majority of steam-generating systems use differential-pressure devices to measure steam flow. A restriction is placed in the flow path—usually an orifice plate —and the pressure difference is measured by a simple, well proven differential-pressure (DP) meter (Fig 51). If needed, the hydraulic output can be converted to electronic signals.

Disadvantages of orifice-type DP meters are non-linearity, higher energy

Fig 51: Differential pressure transmitter

consumption; accuracy can be affected by changes in fluid density and viscosity. They are also affected by impurities in the fluid deposited on the restriction plate. Because they create a pressure drop that is proportional to the square of the flow rate, their useful range is limited to about 3:1.

As the role of energy management receives greater attention, engineers apply alternative flowmeters with more frequency. Categories include: vortex-shedding meters, direct-reading meters, ultrasonic meters, turbine meters, and averaging pitot tubes. Doppler-effect flowmeters, magnetic induction meters, thermal-dispersion meters, and positive-displacement meters are also available, but are either not fully developed for industrial applications or seldom used for steam/condensate measurement.

In reviewing flowmeters, keep in mind the following definitions. Accuracy is the maximum difference between the flowmeter scale reading and the true flow rate. It can be expressed as either a percent of full-scale reading (full-scale accuracy), or a percent of the flow rate at which it is observed (rate accuracy). Operating range, sometimes called turndown, is the ratio of full-scale flow rate to the minimum flow rate at which the meter can achieve its specified accuracy.

Linearity is a measure of how much the meter output deviates from a straight line through the origin and full-scale flow. Non-linearity is the maximum deviation from the line as a percent of full-scale flow.

Repeatability is the ability to give the same output each time the measured flow rate returns to a set value. Reynolds number is a coefficient that correlates fluid viscosity, density, and velocity. Meter calibration factors are given at specific Reynolds numbers. Finally, Newtonian fluids are liquids or gases that behave in a defined physical manner. Flow measurement of non-Newtonian fluids requires special consideration.

VORTEX-SHEDDING FLOWMETERS

These devices operate over a wide range with good accuracy, and can meter steam and condensate flows with no effect from minor changes in density or viscosity. The meter electronically counts a series of alternating vortices that form naturally when a liquid or gas flows around a nonstreamlined object. As the fluid flows around the obstruction, it is unable to follow the shape of the downstream side and breaks away in a vortex. The frequency at which these vortices are shed is proportional to the rate of flow. The meter depends on electronic sensing elements located right at the point of measurement to count the vortex-shedding rate and transmit a digital signal proportional to the flow.

ULTRASONIC FLOWMETERS

When a sound wave is transmitted through a fluid in the direction of flow, its velocity increases; when against the flow, velocity decreases. Measuring these changes is the principle behind the ultrasonic meter (Fig 52).

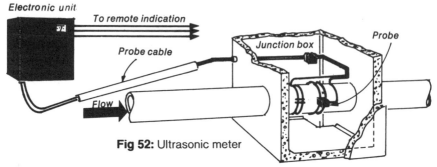

Fig 52: Ultrasonic meter

In the simplest case, two transducers are mounted in the pipeline, and ultrasonic pulses are transmitted from one to the other at an angle through the fluid. If the angle between the sound-wave path and the fluid flow, and the speed of propagation of the sound wave in a still fluid are known, the average fluid velocity can be determined and converted to a volumetric rate of flow.

Output of an ultrasonic meter is influenced by changes in temperature or density of the fluid which affect sound velocity in the still fluid. Also, the fluid-velocity profile must be symmetrical, so the meter must be located well upstream of any pipe bend or obstruction.

DIRECT READING METERS

Some flowmeters not often used for control purposes are frequently used for reading flow. Best-known is the variable-orifice meter, commonly called a rotameter.

Unlike the orifice-plate meter, the flow-restriction size in a rotameter varies with the flow rate to produce a constant differential pressure. The meter consists of a tapered, vertical glass or metal tube containing a float. Fluid flows through the bottom of the tube to the top and moves the float upward. As it rises, the annular space increases until the differential pressure produces a lifting force equal to the weight of the float. Flow rate is indicated by the position of the float.

Useable rangeability of the rotameter is around 10:1. It exhibits a low pressure drop, is easy to install and maintain, and is not affected by upstream or downstream piping configurations. Its range limit in condensate service is near 250 gpm. But its usefulness is restricted by its fairly bulky size and by the glass tube, which will not handle very high temperature and pressure, and gets dirty in some fluids. Steel tubes can overcome some of these problems but it is more complicated to read the position of the float.

TURBINE METERS

Measuring the speed of a rotating turbine mounted in a moving fluid stream provides an accurate and highly repeatable indication of flow rate. Turbine meters (Fig 53) are frequently used when high accuracy is needed under

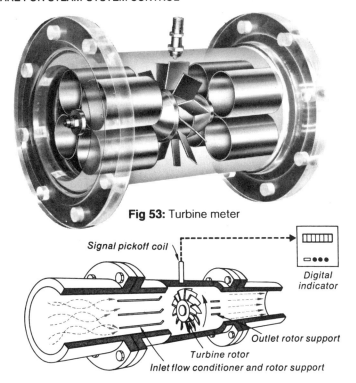

Fig 53: Turbine meter

Signal pickoff coil

Digital indicator

Outlet rotor support

Turbine rotor

Inlet flow conditioner and rotor support

extreme conditions of temperature and pressure, such as in the measurement of high-pressure steam in a supply header. But their cost is relatively high, especially for larger pipelines, and they demand maintenance.

Advances in electronics have enhanced the usefulness of turbine meters. Turbine speed, which is assumed proportional to fluid flow, is read by either a magnetic or a radio-frequency pick-off coil, which senses each turbine blade as it passes the coil.

PITOT TUBES

A pitot tube generates less unrecovered pressure drop than an orifice plate, but conventional designs can measure velocity at only one point in the pipe's cross section, and are liable to clog quickly. Multiple pitot tubes can help, but then multiple pressure transmitters are needed as well.

An averaging pitot is a better solution. Here, the pitot tube is contained within a hollow probe or plenum. Openings in the probe are located so that they can detect differential pressure from equal cross-sectional areas of the pipeline.

The averaging pitot offers about the same accuracy turndown as other DP devices. Unrecovered pressure loss is 10-20% of the differential pressure generated in the pitot tube, compared to an orifice plate where unrecovered pressure loss is typically 60% of the generated differential pressure.

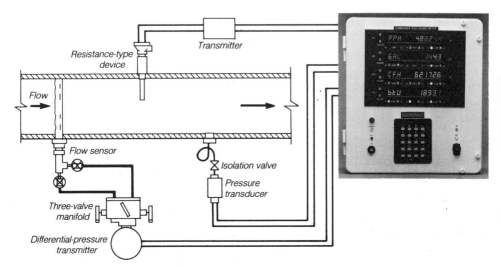

Fig 54: Differential-pressure meter (inset) and system diagram

FLOWMETERS

Once the steam measurements—differential pressure, temperature, etc—have been taken, the data is used to calculate parameters of use to steam-system operators, engineers, and energy managers. Here is where flowmeters enter the picture (Fig 54). Often these meters can interface with all of the primary metering devices described earlier.

Many of meters available today are really micro-computers equipped with digital readout, totalizing and split-second averaging capability, can be programmed for a variety of inputs—instantaneous flowrate and pressure, totalized flow, differential pressure, pipe identification etc—and whose output can be directed to recorders, transmitters, central computers, and data loggers.

Flow-measuring devices are being applied in new ways in today's steam system. To illustrate: Steam-trap operability can be tested by monitoring steam flow, indicating need for trap replacement. Accurate steam-flow monitors can be used to perform control functions for valves, relays, and pumps, thereby saving the costs of additional micro-controllers.

Chapter twelve:
Special topics
concerning valves

There are a variety of design and operating topics—noise control, cavitation, high-pressure-drop service, and fire safety are among them—which concern the selection and specification of valves for steam/condensate systems, but which are too involved and complex to cover adequately in a reference book such as this. Only brief introductions are given here. Further details are available in the appropriate references listed in the back

Noise control

Noise in valves is a result of turbulence introduced to the fluid when the valve produces the permanent head loss that is part of its function. Annoyance to people close by or even damage to human hearing is the major reason why noise control is important in valve/piping design. Unfortunately, the design of steam systems to avoid noise problems requires analysis of many factors. Another problem is that piping and equipment near the valve are also responsible for amplifying and/or radiating noise, so quieting the valve is not always enough. Attempts to acoustically insulate a valve alone have met with little success. Therefore, noise control becomes an important design factor in the valve body and the related system.

Noise is chiefly a topic of concern with control valves that operate for long periods at low opening percentages; block or on/off-type valves create fewer problems. Factors that affect valve noise are flow rate, pressure drop, and atmospheric discharge. Higher flow rates tend to promote higher noise levels. High pressure drop is another common cause of valve noise, especially in atmospheric dump valves and safety valves that discharge heavy flows to atmosphere.

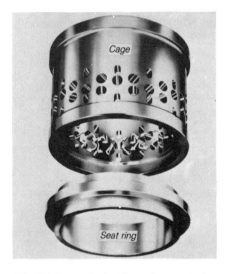

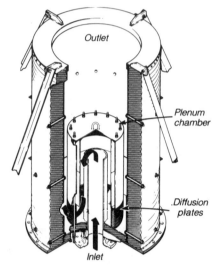

Fig 55: Cage design for noise control   **Fig 56:** Safety valve with silencer

Noise-control remedies are similar to those used to reduce cavitation and wear from high pressure drop, namely, reducing the pressure drop in stages (Fig 55), and/or dividing flow into many small streams. Most of these techniques enlarge the valve, increasing cost.

Devices are available that will attenuate sound by multiple jets, multiple expansions, or both. For example, trim additions for low noise are cages resembling cylinders with many holes. The orifices come either before or after the main orifice. Most control-valve manufacturers provide guidance for predicting and controlling valve noise.

Safety valves cannot have any additional blockage of the discharge, so a silencer (Fig 56) is often used to attenuate sound.

Cavitation

Cavitation damage is caused by the implosion of vapor bubbles contained in the moving fluid against the surface. Evidence of cavitation is the blasting away of the metal surface, which looks as if it had been sandblasted. Even tiny vapor bubbles exhibit forces large enough to damage the metal, if they are close enough to the wall when they are on the verge of collapse and reversion to the liquid state. Stresses from cavitation have been measured as high as several hundred thousands psi.

Three conditions are required for cavitation to occur. First, the valve must be handling a liquid that is able to vaporize or flash into bubbles somewhere in the valve flow path. If the liquid has flashed upstream, the fluid will be two-phase, with large vapor bubbles in the liquid and/or possibly even liquid bubbles in the vapor.

Second, flashing of the fluid must take place, though flashing itself does not cause cavitation. If the pressure drop across the valve transforms enough static pressure into dynamic pressure, then the static pressure can go below the saturation pressure for the liquid at the temperature in the orifice.

Third, bubbles must collapse for cavitation damage to occur, and fourth, the bubbles must collapse at the metal surface. If downstream static pressure exceeds the saturation value, bubbles will collapse.

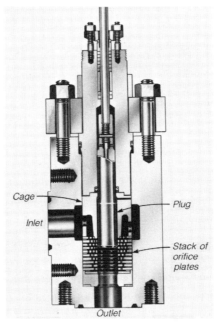

Fig 57: Multistaging to prevent cavitation

Cage

Inlet

Plug

Stack of orifice plates

Outlet

Multistaging is one remedy (Fig 57), as long as the static pressure in all the stages stays above the vapor pressure. Some valves with inherently high pressure drops are less likely to cavitate. Special discs or cages are other protection techniques, but add costs. Today's high-technology control-valve trim often combines multiple flow paths and tortuous passages with multistaging to prevent flashing and subsequent cavitation (Fig 58).

High-pressure drop

As shown in another chapter, high pressure drop across a valve is encountered in boiler-feedwater-flow control, boiler-feedpump minimum-flow recirculation, steam-bypass and dump service, and boiler-blowdown service. Actually, any valve on a high-pressure line can experience high pressure drops, at least for short periods of time.

Sometimes even more damaging although less suspected is the high pres-

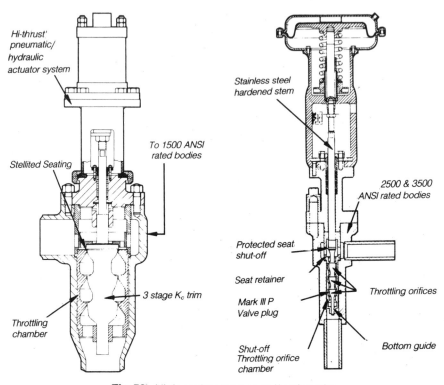

Fig 58: High-technology control-valve trim

sure drop that occurs across closed valves, if there are leakage paths under a high differential pressure.   Result of high pressure drop unaccounted for in the design stage is localized and broad cavitation damage and erosion, especially if the valve handles a liquid that contains solids.

Design remedies for the effects of high pressure drops are multistaging of the pressure drop, dividing the flow into many small flow streams following tortuous paths, or containing the jet energy by centering flow from many jets on a point inside a cage or downstream of the orifice zone, or restrictions downstream of the valve. Attempts have been made to use two or more valves in one line to break the pressure drop without exceeding the limits of either one. But this is not only an expensive option, but one that has failed in many instances due to problems in balancing the pressure drop through both valves over a range of flows.

Several valves designed for high pressure drop have cages of stacked plates with labyrinths, involving many turns and repeated enlargements and contractions, that produce small low-energy jets. In others, the flow passes instead from plate to plate through orifices. Still other designs utilize multiple throttling contours on a common stem operating in a throttling chamber to break the total pressure drop into smaller stages.

In high-pressure-drop work, the effect of leakage past close-fitting surfaces or seal rings may eventually ruin the valve.

Of help in high-pressure-drop applications is a replaceable orifice downstream of the main orifice which builds pressure at the valve outlet and reduces the potential for cavitation. Low-flow conditions can injure the main valves, but during high flow, the orifice can be advantageous.

Fire safety

Valve/actuator operation during a fire is a vital concern not only for petroleum and chemical plants but also for powerhouses where oil, gas, or other flammables such as waste streams are burned to generate steam. Greater use of soft seats in valves adds to this concern, as does the wider use of automation for valves and controllers. Floating-ball valves with TFE seat rings and concentric butterfly valves with spherical seat contact and TFE seat rings are among those causing the most concern.

Whenever a fire-caused valve failure would increase damage or jeopardizes life and limb, the systems designer must analyze valve action and specify for safe shutdown or opening of lines.

Fire safety extends beyond the valve itself. The actuator may have to resist fire for at least a short period of time, even if it cannot be made as durable as the valve itself. Single actuation may not be enough: In certain circumstances, the valve may have to be actuated several times during or directly after a fire.

Enclosures for actuators have a certain period of fire resistance built into them, which assumes that the external control system is also fire-resistant. These assumptions include not only the air lines, but also many small valves, wiring, and relays. Constant inspection is necessary if the enclosure is to be ready to resist fire at all times.

SECTION V:
STEAM-SYSTEM-CONTROL HARDWARE APPLICATIONS

Chapter thirteen:
Important valve and
controller services

Familiarity with the available hardware for the control of steam/condensate systems is a first step towards proper selection of equipment. Before hardware can be specified or purchased, the operating environment in which it will serve must be properly characterized in terms of flow, pressure, temperature, and level ranges, in terms of the composition of the fluid, and in terms of the normal, off-design, and emergency operating procedures.

Many of the steam/condensate-handling and control components are fairly standard and cost-effective and have been proven in particular services throughout years and years of reliable operation. Others are not straightforward at all. The criticality of the service to the overall system, the severity of the environment, and/or the difficulty of the operating constraints all can make selection of hardware complex. Usually, the complexity stems from matching the hardware to very rigid specifications, or from achieving a reliability and operating life that are acceptable to the user.

As the importance of steam handling grows out of a generally heightened awareness of energy costs, more valve and controller services are elevated to important or critical classifications. Consider steam traps. Not many years ago, most US plant personnel neglected these lowly items in maintenance programs because the conservation of steam and condensate was of little importance. Today, facilities with many traps have maintenance programs specifically devoted to them.

Many important valve and controller services are common to all steam systems, large or small (Figure 1): boiler-feedwater control, main-steam control, boiler-blowdown control, feedwater deaeration, and steam-drum-

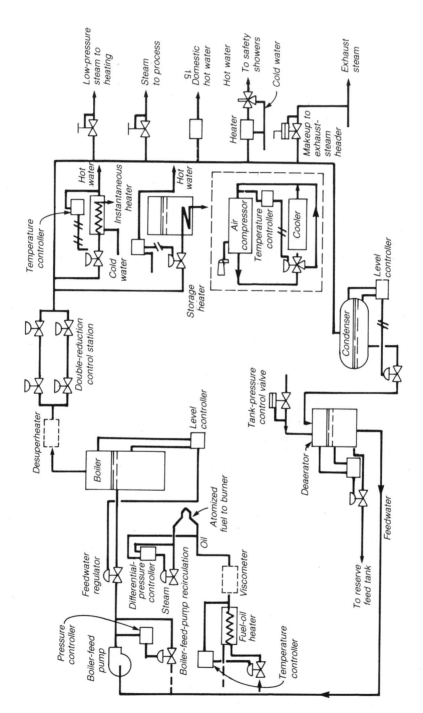

Fig 1: Typical steam/condensate system

pressure control. Industrial and commercial steam systems have added important services associated with transporting steam, reducing its pressure, and returning condensate through, in some cases, miles of complex piping systems. Steam-pressure reduction, turbine-backpressure control, and condensate-return and drain lines are some of the services more peculiar to these smaller systems. Turbine-inlet and extraction-steam control and desuperheater flow control are two added services important to large industrial and utility plants.

Higher temperatures and pressures and the objective of producing power as efficiently and reliably as possible force several other important services on utility steam systems. These services include turbine-bypass control, and, in some utility boilers, recirculation. And, of course, nuclear steam systems are in a class by themselves due to strict codes and regulations for construction. Here, feedwater bypass, main-steam isolation and feedwater isolation, relief valves, and steam-dump valves are among the services that have yet to be satisfactorily proven.

The rest of this chapter is devoted to describing many of these important/critical services with respect to the operating constraints, pressures, temperatures, and flow and level characteristics of the fluid as well as the fluid composition.

Feedwater control

Accurately controlling feedwater to the boiler is a basic and critical steam-control service, especially in modern, high-pressure boilers. Usually, the feedwater regulator follows a constant-pressure feedpump (Figure 2). Normal operating conditions are not severe. But during startup and shutdown,

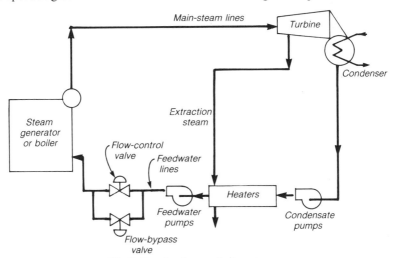

Fig 2: Feedwater control system

there are complications resulting from rising feedwater-pump-discharge pressure combined with low boiler pressure. This situation causes abnormally high pressure drops across the valve. If the valve operates for prolonged periods under this condition, valve-plug vibration and/or cavitation can result.

The feedwater-flow-control device must be compatible with the combustion-control equipment as well. Pneumatic, piston-type actuators and the electrohydraulic piston-type actuator are the most commonly used power sources for positioning the valve plug. When large valves with high mass velocities are encountered, such as with feedwater control valves, the designer should consider providing the control device with a hydraulic snubber.

Centrifugal boiler feedpumps require a minimum flow through them to carry away the pump's drive energy; otherwise, the steam provided by it would ruin the pump. Ensuring this minimum flow and preventing low-flow instability in feedwater-recirculation lines needs further investigation.

Usually in powerplants, a recirculation line from the pump to the deaerator or condenser serves this purpose (Figure 3). This line can be quite long,

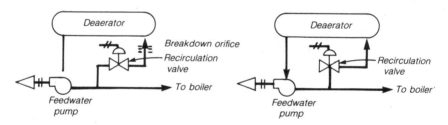

Fig 3: Feedwater recirculation line

with considerable pressure drop at high flowrates, causing backpressure on the valve outlet. The conventional method of control is to insert in this line a breakdown orifice in series with a control valve to draw off up to 50% of the rated flow. The control valve acts in an on/off mode for a given flow rate through the feedpump. Valve pressure drop is set at a level below the cavitation range and the rest of the drop is taken across the orifice. This method has been somewhat successful for smaller low-pressure pumps. Tight shutoff of the valve is mandatory because of the high pressure drops; even minute leakage can destroy the valve trim.

As pressure and temperature of the feedwater increase, more recirculation flow is necessary. For example, nuclear and supercritical fossil-plant feedwater pumps have recirculation percentages as high as 50% of pump capacity. To avoid multiple-valve installations and complicated controls, modulation is essential in the recirculation line. The problem is achieving this modulation without many hours of operation at partial valve opening, thus inviting cavitation, vibration, and other problems. In earlier days, valves would last anywhere from several hours to a year at most. New valve designs and installation procedures have met this challenge. Three principal design concepts for these valves are multi-staging, tortuous paths, and multiple orifices.

Main-steam control

Controlling the flow and pressure of the main steam exiting from the boiler was relatively straightforward--the majority of boilers operated at constant steam pressure--until the pressures and temperatures typical of large utility boilers came into play. Low-load operation also aggravates pressure control.

Several US utilities operate supercritical steam generators. Pressures in such units are above 3200 psig—near the value where the specific volume of steam and water are equal. Steam conditions from a once-through supercritical unit are easier to control than from a drum-type unit. An increase in feedwater flow results in an immediate increase in steam flow. Also, steam temperature can be controlled more accurately because the inlet temperature to the superheater is not fixed.

Some utilities have modified their subcritical forced-circulation units so that they operate as once-through supercritical ones above a given percentage of the full load. Existing supercritical designs use boiler throttle valves at the waterwall outlet to maintain pressure in the waterwalls. Above a certain load, turbine-load control must be accomplished using the throttle valves; this results in large losses that lower efficiency and cause wear on the turbine valves.

Operating the boiler under varying pressure is one way to eliminate throttling, but the steam-pressure and control aspects are more complex. For one thing, mass flow in the tubes must be high enough to avoid excessive metal temperatures, especially at supercritical pressures.

Steam-temperature control

By far, the most popular method of controlling steam temperature from a boiler or other exchanger is with a desuperheater. Three common locations for desuperheaters are downstream of steam-pressure-reducing valves, in steam-turbine-extraction lines, and following boiler superheaters and reheaters (Figure 4). For utility powerplants, the major objective is to hold steam temperature within the constraints of the piping materials. Industrial plants use desuperheaters additionally to fulfill the steam-temperature requirements of a particular process or unit operation.

Most desuperheaters are fundamentally similar. Cold water is brought into contact with superheated steam. As the steam gives up heat to evaporate the water, its temperature drops. The final steam temperature is a direct

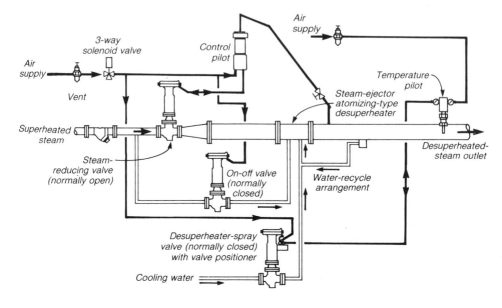

Fig 4: Desuperheater control station

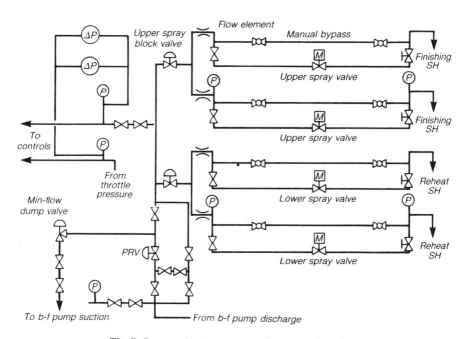

Fig 5: Desuperheater spray-valve control system

function of the amount of water injected. Many different types of desuperheaters are available to perform this job.

Desuperheater spray valves can be a source of costly problems when low flow rates and high pressure drops work in combination (Figure 5). High pressure drops result from the difference between the feedpump discharge pressure and the superheater pressure. For example, startup conditions often result in low boiler pressure which adds to the pressure drop across the valves. Correspondingly, flow through the valves during startup can be less than 10% of the maximum.

Transients complicate matters. During load pickup or reduction, spray water flow will vary considerably. Therefore, throttle control is mandatory with desuperheater-spray control valves. Throttling control of low flow at high pressure drops dictates special valve-trim considerations. To illustrate: Cage-guided trim valves may not be adequate, because they have more clearance flow than needed in this service.

The spray valve of the desuperheater in the primary superheat section must have as its inlet pressure the full discharge pressure of the feedwater pumps. However, the reheat section involves lower steam pressures. One recommendation is to tap an inner stage of the boiler feedwater pump where the outlet pressure is always higher than the reheat-section pressure. This option lowers the pressure drop across the spray valve for the reheat-section desuperheater.

Desuperheaters require tight shutoff of the spray valves for the same reasons as the feedwater-recirculation control valve: potentially high pressure drops.

Some powerplants have performed extensive review of desuperheater systems to avoid what they have experienced as perpetual problems with the spray valves. In some cases, the analysis has unlocked design considerations other than the valves themselves.

Safety and pressure relief

One valve function common to all steam systems is pressure relief. Valves for this service are designed to protect fluid-handling systems from excessive overpressure. They must open automatically at excessive rises in pressure, then close promptly and be leak-tight when normal pressure is restored. The American Society of Mechanical Engineers (ASME) Boiler and Pressure Vessel Code Committee has developed operating guidelines for this service.

For steam-heating boilers, the pressure differential between the safety-valve set pressure and the boiler operating pressure should be at least 5 psi. Remember that the recommended maximum operating pressure for a steam-heating boiler is 10 psig. If the boiler operating pressure is greater than 10 psig, the differential should not exceed 15 psig minus the valve blowdown pressure. For power boilers that operate between 300 and 1000 psig, the differential is 7% of the operating pressure, with 30 psi as a minimum; between 1000 - 2000 psig, the figure is 5% with 70 psi as the minimum. Over 2000 psig, the differential is left to the discretion of the designer. (Remember that when dealing with engineering codes, be sure to reference the latest set of revisions.)

Care should be taken not to operate a boiler at or near the set relief pressure. At pressure near the set point, a safety valve tends to weep or simmer (leakage), causing deposits to accumulate in the seat or disc area. This condition has been known to cause a valve to freeze in the closed position, and fail to open prior to reaching the rupture pressure of the boiler or pressure vessel.

Solids present in the steam and feedwater aggravates this problem, even if the safety valve has never opened. Minute leakage through the seat carries solids out of the valve and into close-tolerance guide clearances. Individual valve manufacturers have their own design modifications to prevent simmering and effects of deposit accumulation.

These modifications tend to be more extensive for the modern high-pressure boiler. In fact, overpressure protection of utility steam generators has evolved into a hybrid system that relies greatly on elements other than spring-loaded safety valves. One design incorporates a power-actuated valve in which the disc closes upward against a seat, so that the boiler pressure tends to increase the seating force and reduce the chance for leakage. This type of pilot actuation offers the ability to set the opening pressure within close tolerances, but reliability of all the valve components is paramount.

Boiler blowdown

Impurities in the form of suspended and dissolved solids are left behind in the boiler drum when the steam is formed. Some of the solids settle in the lowest parts of the boiler while others tend to float on the surface of the drum water. An intermittent or continuous blowdown system must be designed to remove these impurities. Continuous blowdown—while more popular for steam generators with a large amount of makeup water—improves plant performance if a flash tank is used to recover low-pressure steam, perhaps for deaeration, and conserve expensive condensate (Figure 6). Valves and controllers for blowdown systems represent a particularly difficult service. Part of the difficulty is due to the erosive nature of the solids. Another part stems from the inherent two-phase flow that results when the blowdown water flashes.

In a typical continuous blowdown system, the steam flashes in a tank which is slightly above the deaerator pressure, or the pressure of whatever equipment is fed by the flash steam. A control valve in the drain piping to the heat exchanger or blowdown tank maintains condensate level in the flash tank. Condensate, flashing at the control valve, flows to the blowdown tank,

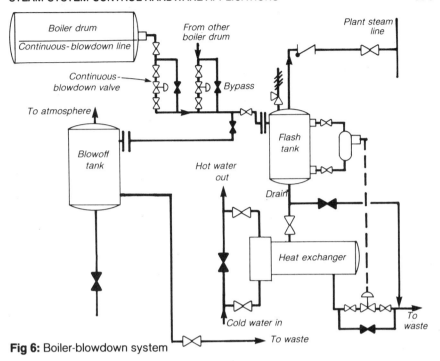

Fig 6: Boiler-blowdown system

which is at slightly above atmospheric pressure.

Valve choice and configuration depend on the boiler operating pressure. Commonly, those over 100 psig require two blowdown valves in tandem—either two slow-opening valves or one slow-opening and one quick-opening valve. One acts as the sealing valve, the other as the blowing valve. For pressures up to 600 psig, a tandem arrangement of a hard-seat blowing valve and a seatless sealing valve can be used. For higher pressures, the hard-seat sealing valve is used. Some blowdown valves come as a one-piece body that contains both the sealing and blowoff valves.   In all blowdown valves, both valves should be either fully opened or fully closed to prevent erosion of seat and disc faces and increase the life of the packing and working parts.

Feedwater-heater drain valves

As explained in chapter five, feedwater heaters are exchangers where steam, extracted from various locations within the turbine, is used to preheat boiler feedwater. A modern steam powerplant will have anywhere from one to over twenty feedwater heaters, depending on the cycle and the size of the plant.

Some feedwater heaters are of the direct-contact variety. That is, the steam and feedwater are combined to raise the temperature of the mixture. However, most are closed, or surface heaters. Here, extracted steam condenses on the shell side and transfers latent and sensible heat to the feed-

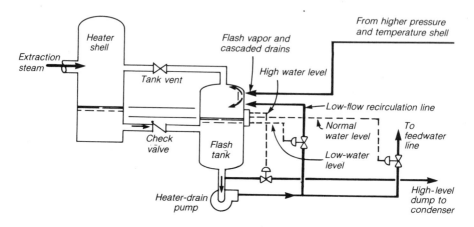

Fig 7: Feedwater-heater level control

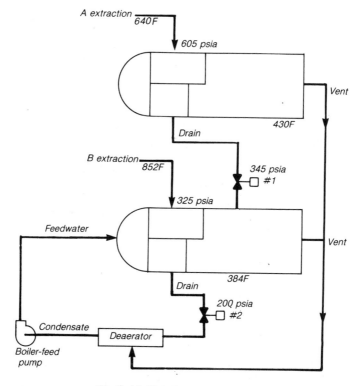

Fig 8: Multiple-heater system

water flowing inside the tubes. Feedwater heaters usually come equipped with two control valves discharging into separate locations, one to handle startup and emergency low- and high-level situations and another for normal level control.

Condensate on the shell side is usually removed automatically and continuously through a level-control valve (Figure 7). This service is critical because, in many plants, the heaters are arranged in series or in parallel, and each one is dependent on the stable level control of the one preceding it. Second, the water level in the shell affects the thermal performance of the heater. Third, condensate can, in extreme instances, back up into the turbine, backflood into desuperheaters or condensation zones, and promote water induction into the turbine.

Consider the feedwater-heater configuration in Figure 8. Operation of either drain valve affects the flow through the other valves. For example, if flow through valve one is increased, extraction flow B decreases and extraction flow A increases. The position of these valves regulates the condensate level in the shell, and in turn, the pressure drop. Low water level creates pressure drops that cause higher than desired steam velocities, harmful flashing conditions, and subsequent erosion of the tubes. High water levels reduce the amount of surface area available to condense steam.

Sizing the control valves and piping for the expected flashing conditions is the problem here. By design, the heater drain valve will handle subcooled condensate which can flash in the valve body as its pressure is reduced. Flash calculations normally assume no flashing upstream of the valve. But since this has a high probability of occurring, stability problems result as the valve tries to adjust to very different flow requirements of the steam and condensate phases.

Common body styles for heater drain valves are: angle construction with a venturi-type liner and steel body, globe construction with an erosion-resistant alloy body material, or throttling-type ball valve constructed of 316 ss. This last selection is particularly adaptable to the larger low-pressure heater drains.

Deaerator-level control

Deaerators are open direct-contact heaters which use the heat from low-pressure steam to release most of the noncondensable gases contained in the boiler feedwater, especially O_2 and CO_2 for corrosion prevention. The deaerator is normally situated between the low-pressure feedwater heater and the high-pressure feedwater heater in utility systems. In smaller industrial steam systems, it may be the only piece of equipment—not including the pump—between the condenser and the boiler.

Level control in deaerators involves high pressure drops at low flows and very low pressure drops at high flow rates (Figure 9). Added pressure differ-

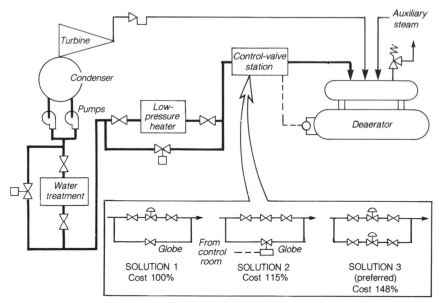

Fig 9: Deaerator level control

entials occur due to the drop across the water-treatment beds and the heaters, if they are fouled or out of service.

This service dictates a valve with high rangeability. Ball valves come to mind first, but are susceptible to cavitation damage at low-load conditions. One possible way out of this is to insert a tube bundle in the curvature of the valve ball to create backpressure, relieving the drop taken between the tube bundle and the ball seal, and to prevent formation of a severe vena contracta. Cage-guided valves, with a trim designed for cavitation resistance at low flow rates, or valves with high C_v coefficients, are other possibilities.

Another alternative is to operate two control valves in parallel. Each is sized to control the system during normal operation, meaning low or medium pressure drop in the water-treatment system, heaters, etc. This scheme also offers higher plant reliability because of the backup valve—but the cost increase is obvious.

Turbine-bypass systems

Turbine bypass is being applied to older drum-type fossil-fired powerplants, originally designed for baseload service, to help them handle the rigors of increased cycling duty required of today's powerplants. Objective of bypass systems (Figure 10) is to reduce the time and cost of daily startups by matching steam temperatures with temperatures of the turbine rotor and casing, thereby avoiding large thermal stresses. Bypassing steam to the condenser allows the boiler to reach the required operating conditions before the

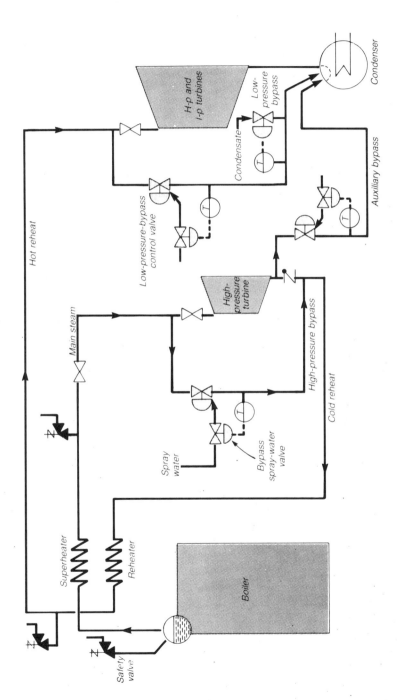

Fig 10: Turbine-bypass system

turbine is rolled. They handle anywhere from a few percent to 100% of the steam load.

Other advantages include: reduction of the solid-particle erosion damage to high-pressure turbine blades, ability to decouple the boiler and turbine, and ability to avoid boiler trip following a full electric-power-load rejection. Depending on applicable boiler codes, high-pressure bypass systems can replace relief and safety valves.

High-pressure bypass systems direct steam from the secondary super-heater outlet to the cold-reheat line through a pressure-controlled valve. Desuperheating is accomplished along the way. To protect the h-p turbine against reverse flow from the cold-reheat line, a positive-closing check valve is installed in the cold-reheat line. In the low-pressure version, hot reheat steam flows to the condenser whenever the reheat steam flow exceeds the demands of the i-p and l-p turbines. To protect the condenser, a two-position stop valve, closed during normal operations, can be added. When the low-pressure system is in use, the stop valve opens immediately, leaving the bypass valve in command.

Valves for high-pressure systems have been adapted from the bypass and desuperheating needs of once-through boilers. In the once-through power-plants, bypass valves are arranged so steam pressure causes them to fail open. Drum-boiler plants require fail-shut valve actuation. One way to achieve this is to mount an emergency hydraulic accumulator close to the valve actuator. It holds the bypass valve shut should a loss-of-control signal or complete failure of the normal control hydraulics occur. Other valves have been modified to accommodate the fail-shut needs of drum plants.

Nuclear-plant services

Special valve services in nuclear powerplants include steam dump or bypass, and main-steam and feedwater isolation (Figure 11). In addition, valves for more typical duties are subjected to more rigorous constraints than valves in fossil-fired powerplants.

In pressurized-water reactors, an atmospheric steam dump and turbine bypass are part of the steam cycle. During startup, steam is diverted from the turbine to the condenser, used as a heat sink for the reactor. If an emergency occurs, forcing rapid shutdown, steam is diverted to atmosphere.

Both services demand similar valve characteristics. Steam flow must be precisely modulated and throttled, the valves are required to open in three seconds or less, and because they remain closed most of the time, must exhibit extremely tight shutoff. Pressure drops and temperatures are as high as 1000 psi and 560F.

Isolation valves—used to ensure safe handling of potentially radioactive fluid—also must close in the 3-5 second range, and provide uncontested tight shutoff. Other desirable characteristics include: absence of small moveable

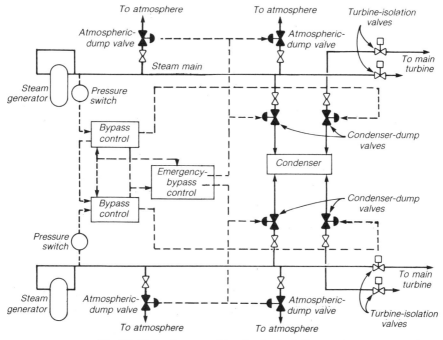

Fig 11: Main-steam and feedwater isolation system

internal parts and body penetrations, low valve-induced pressure drops, and uniform and moderate actuation force. Since isolation valves require no throttling capability, the gate valve is a logical choice for this service. Globe valves have also been used, but they must be configured to minimize pressure drop—such as with a Y-configuration. Their use entails precautions concerning internal/external springs and/or the control piping. Ball valves are another good choice as long as the seating has been designed properly for this service.

Isolation valves cannot just slam shut within a few seconds. The closure must be controlled to prevent excess flow and protect the valve. Recently, the requirement to close against flows in both directions has been added. A balanced globe valve with spring actuation can meet the requirements. If that isn't enough, the force of steam itself can power the valve by use of an accumulator, sometimes supplied as part of the actuator. Accumulators close main-steam isolation valves much faster than, say, an electric-motor-powered actuator, and reduce the chances for damage to the valve.

Operability of many valves, especially check valves, must be determined on line. Use of transducers is one way to detect the position of a valve, and is a common design feature of nuclear-plant valves. Safety and reliability require that valves, like the rest of the plant, undergo vibratory and seismic analysis and analysis for the effect of radiation on metals and non-metals. Often, the results will add constraints to the size, weight, and mounting of actuators and eliminate the use of certain common materials such as Teflon in certain services.

Valves for nuclear plants demand higher thrust, speed, and resistance to environmental factors compared to other services. Here's an example of the type of problem: To satisfy closing-speed requirements, speed of electric-motor actuators was increased by simple gearing changes. Unfortunately, the inertial effects of the high-speed motor can damage the valve as the disc approaches its seat. This effect is most evident on valves with rigidly mounted seats to help resist leakage.

Control valves for nuclear service have notable differences in design criteria from other applications: 1) The number of valve operating cycles is greater over the life of the plant; 2) they must have provisions to contain possible radioactive leakage. As an example of the effect of the latter, modulating valves usually have two sets of stem packing, with a leakoff between them that feeds a low-pressure waste-collection system.

The myriad of nuclear-plant construction and design codes and standards— the American Society of Mechanical Engineers (ASME), American National Standards Institute (ANSI), and American Society for Testing and Materials (ASTM) are three bodies issuing codes—must be thoroughly reviewed before materials of construction and welding and installation procedures can be established for valves for nuclear plants.

Several aspects of nuclear safety valves are worth singling out. For one, nuclear plants have spurred the use of larger and larger safety valves that can discharge over 2 million lb/hr of steam in the 200-300 psig range. Moisture separator/reheaters, deaerators, and turbine-exhaust lines are among the areas safety valves protect. The enclosed-spring design is common but a pilot-operated valve is also used. A small spring-loaded safety valve opens and delivers line pressure to a shuttle valve that then opens and relieves line pressure from above a piston-like disc of the main valve. Line pressure above the main disc assures tightness.

Unlike most fossil applications, nuclear safety valves often do not discharge to atmosphere, but to a line that can have varying values of backpressure. To prevent this backpressure from affecting the opening pressure, a bellows can be used that surrounds the spindle above the disc holder. An auxiliary benefit is that it prevents escape of fluid to the outside or into a closed bonnet.

Pressure reduction

As noted in earlier chapters, steam at many industrial and institutional facilities is generated at the highest temperatures and pressures required of the process or mechanical design, then reduced in pressure to meet the needs of individual steam users. Here, pressure-reducing valves (PRV) find wide application.

Pressure reduction can be a difficult service for a variety of reasons: range of flow and pressure-drop conditions that must be handled by the reducing station, possible erosion caused by high velocity, contaminants, cavitation, alone or in combination, and, finally, noise.

Wide flow and pressure-drop ranges may be due to the variations between the normal and emergency operating conditions—such as in refineries where the plant runs at steady state for months at a time until an emergency or scheduled shutdown occurs. Or the wide range may occur because of normal seasonal variations. Steam-piping systems for space heating of commercial buildings are an example of normally wide-ranging flow conditions. Average winter steam demand can sometimes be less than half of the design steam load. In addition, steam for warming up the system rapidly—to match metal temperatures to that of the steam—may exceed the design operating load.

As steam travels through piping and vessels in industrial complexes, it picks up solid and liquid contaminants. These erode valve components in the fluid path. Erosion is aggravated at higher velocities.

There are two common classes of service for PRVs. One is where the steam flow must be completely shut to prevent buildup of pressure on the low-pressure side. Another is where the low-pressure lines will condense enough steam to prevent buildup of pressure through normal leakage through the valve (some types of PRVs do not provide tight shutoff). For steam-heating systems, the American Society of Heating, Refrigeration, and Air Conditioning Engineers (ASHRAE) recommends single-seated valves, either direct-operated or pilot-controlled for the first class of PRV. For the second, it recommends direct-operated or pilot-operated double-seated valves. Double-seated valves are less affected by variable inlet pressure, while pilot-controlled single- or double-seated valves help eliminate the effects altogether.

For years, single- or double-seated globe valves served as pressure-reducing valves. To overcome the noise and erosion prevalent with this simple design, cage-type-trim valves were introduced to this application. Most of these valves work on a common principle. A plug moves up and down inside a drilled-hole or slotted cage, and steam passes through the openings to the valve outlet.

More recently, valve designs that accomplish staged pressure reduction, and the use of outlet diffuser plates have helped control steam velocity and noise and extended the service life of pressure-reducing-valve internal components.

Where there is considerable load variation, such as in building steam heating systems, it can be good practice to install one large and one small valve in parallel (Figure 12), then use whichever one fits the load best at a given time.

Hand-operated bypass valves are essential for most pressure-reducing stations, so the PRV can be serviced without interrupting operation.

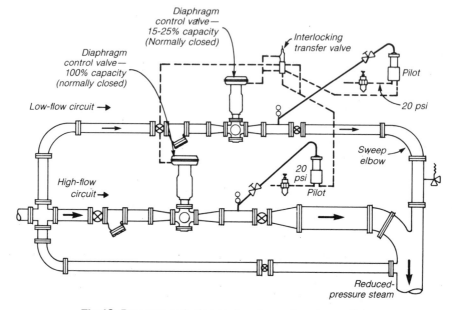

Fig 12: Pressure-reducing station, two valves in parallel

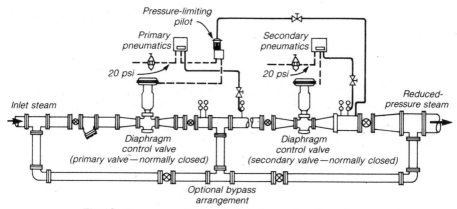

Fig 13: Pressure-reducing station, two valves in series

Sometimes, use of two reducing valves in series (Figure 13) is a way to maintain reduction stability and improve safety while accommodating wide ranges of flow and pressure drop. The second valve, set at a slightly higher reduced pressure, is normally wide open and only throttles the flow of high-pressure steam resulting from failure of the primary valve. Distance between the valves in the pipe is an important criterion to reduce excessive "hunting" by the upstream valve.

Pressure regulators, in contrast to control valves, are also available for many pressure-reducing duties. They offer near-instantaneous response because the process fluid is in contact with the diaphragm. Maintenance is simple because they contain fewer parts, and they require little or no external piping, being self-contained.

High-pressure letdown stations often require desuperheaters to maintain appropriate steam temperatures. For direct-contact superheaters, the water-spray nozzle is either located downstream of the PRV or is incorporated into the PRV, so that water evaporates within the valve body. One advantage of the latter is that it handles capacity turndown better because steam velocity and flow turbulence are relatively high at the water-injection point. Other potential advantages include lower sound levels and elimination of long straight-piping runs. With a combination valve, temperature sensors can be placed the proper number of pipe diameters downstream of the valve.

Condensate lines

Valves on condensate and return lines—such as around steam traps—represent a very difficult service, especially for industrial steam systems which can include hundreds of these valves (Figure 14). Main reasons for the difficulty of service are the two-phase, or alternating-phase fluid flow which delivers water drops or slugs with the steam, flashing water, and varying temperatures in rapid sequence. These effects are aggravated at higher pressures and when feedwater treatment is lacking, causing dirt and scale dislodged in the piping to collect in traps and auxiliary piping and valving. Result? Valves that leak when the component they are isolating requires maintenance.

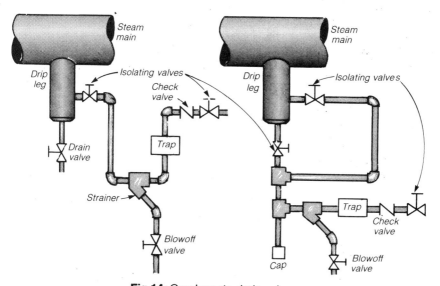

Fig 14: Condensate-drain valves

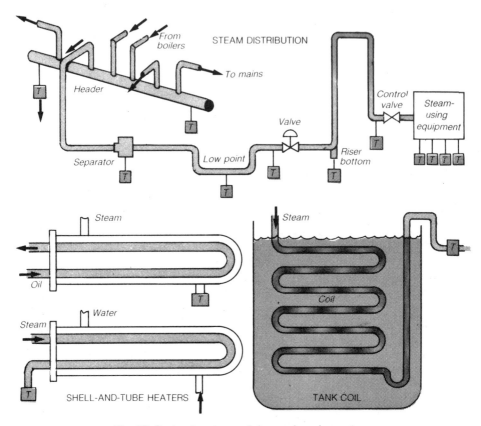

Fig 15: Drains for steam piping and equipment

A general rule for drips and drains is to install a steam trap at every low point and directly upstream of control and isolating valves and valves that can be closed when steam is flowing (Figure 15). If straight-run distances between low points on a distribution line exceed about six hundred feet, intermediate drip legs and traps should be set at 200-500 ft intervals. If the distribution line can be fed from both ends, such as a header, extra trap installations may be necessary.

Steam traps usually require an upstream strainer to prevent dirt and scale from plugging the trap orifice. Thermostatic steam traps normally do not require an upstream strainer. If maintenance will be necessary during plant operation, blocking gate valves should be installed on each side of the trap and the station should include a valved bypass. Bypasses can be used during startup and shutdown for main steam and distribution piping, but more importantly, their use prevents damage from a failed-closed trap. When traps discharge into a common header, installation of check valves is good practice to prevent reverse flow.

Shutoff and check valves often give such poor service that some plant engineers are in favor of eliminating them wherever possible. This situation has to be evaluated site specifically. For example, some bypass valves can be eliminated if there is a steam shutoff ahead of a heater or other piece of process equipment that is in less severe service. It can be used to isolate the trap. Another consideration: If the bypass isn't there, then neither is the potential loss from a bypassed, neglected, or failed-closed trap.

Often the problem can be solved by purchasing better quality valves the first time around. As mentioned at the beginning of this chapter, these components have only recently begun receiving adequate attention from designers, operators, and maintenance crews.

Valves for steam-turbine service

There are a variety of important valve services for steam turbines; the number multiplies when turbine load is to be controlled automatically or when the turbine is equipped with one to several extraction points. In addition to the main-steam inlet valves, turbines have one or more of the following: main-steam stop valves, main-steam stop bypass valves, extraction-steam control valves, backpressure control valve, diaphragm relief valve, exhaust-pressure relief valves, valves for the turbine shaft-packing steam-seal system, and steam-seal blowdown valves. Large steam turbines for utility and industrial duties additionally have reheat desuperheating spray valves (discussed earlier), reheat stop valves, and reheat intercept valves. Most of these are supplied with the steam turbine.

Small industrial steam turbines represent the minimum for steam control. A speed governor, overspeed trip, throttle-pressure gage, thermometer, exhaust-pressure gage, and tachometer are about all that's needed. If the turbine exhaust affects the pressure in the steam header it dumps into, then the governor includes an exhaust-pressure regulator.

Many single-stage steam turbines employ hand-operated nozzle control valves to maximize efficiency at part loads, to achieve rated load at reduced steam pressure, and/or to operate at overload capacity. Steam, as it leaves the main governor valve, fills a steam chamber supplying a nozzle ring. Between the steam chamber and nozzle rings are valved parts that feed steam to certain nozzles or groups of nozzles. Each of the valves can be full opened or closed to adjust the nozzle area so it conforms as closely as possible to the area required by the steam flow at a given load. This type of operation reduces throttling.

It is important to note that these valves cannot be throttled. Doing so will erode the seat rapidly, a condition better known as wire-drawing. Also, these valves should not be closed tightly until all turbine parts reach operating temperature. Otherwise, they leak due to differences in thermal expansion between the valve stem and the turbine casing.

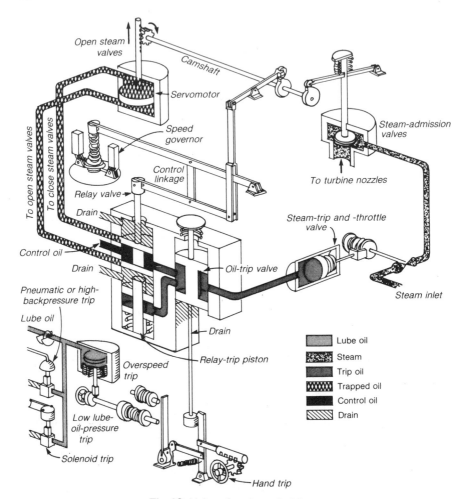

Fig 16: Valves for steam turbines

For a typical turbine design, the inlet-steam-valve casing will contain the main-steam stop valves that feed several control valves. Steam is fed into the valve casing, passes through the main valve and control valve, then is guided through pipes leading to the nozzles on the first stage (Figure 16).

Usually, the main stop valves have one function—to cut off the steam supply in case of a turbine trip. They have no control capability, so they absorb only a slight pressure loss in operation. Design considerations for main stop valves include: tight sealing, resistance to erosion and mechanical damage, protection against inlet-steam pressures when the valve is closed, and bypass valves, used to prewarm the valve chest.

Multiple steam-inlet control valves ensure that as little throttling takes place as possible at part loads. In some cases, the valves' seats are designed as a diffuser with large flow cross sections to minimize pressure drop under unthrottled conditions. Other favorable characteristics for this service include: fast response times with small actuating forces, tight sealing, and ease of testing while the valve is in place.

For internally controlled extractions from a multi-stage turbine, extraction valves control the amount of steam entering succeeding stages of the turbine, keeping the extraction pressure constant. Extraction control is more critical for industrial turbines because steam pressure is a function of the process operation it serves. Utility plants normally have uncontrolled turbine extractions, since the overriding objective is to generate kilowatts. Design features important to the steam-inlet control valves are important here as well.

Automatic constant-pressure extraction requires a control system that regulates both the inlet-steam and extraction-steam control valves. It must position both sets of valves under variations in horsepower and extraction-steam system demand—while maintaining the set governor speed and extraction pressure.

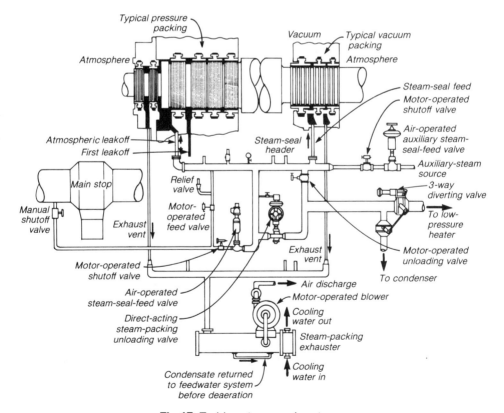

Fig 17: Turbine steam-seal system

A popular way to accomplish this is to use a pneumatic controller as a speed governor in place of the conventional mechanical-hydraulic, or flywheel, governor, and another pneumatic controller for extraction-pressure control. Outputs of each controller are then converted to position signals and combined to meet the flow and power characteristics of the turbine in question. Mechanical linkages—traditionally used to effect this control— have given way to pneumatic relays and other devices with electronics to position the steam valves.

Steam-seal systems (Figure 17) protect turbines and, by extension, the balance of the plant from air leakage, and prevent steam leakage into the turbine-shell surroundings. They become complicated for utility-sized units and involve multiple discharge or drain points such as to the main condenser, steam-seal condenser, or feedwater heaters. Often the discharge point depends on operating conditions, and diverting valves, bypass valves, or relief valves may be required.

Fossil-fired and nuclear reheat-cycle powerplants use stop and intercept valves in the reheat lines to and from the turbine. They are located as close as possible to the turbine-inlet openings to reduce effects of a potential over-speed caused by entrained steam. Both valves are opened during normal operation. The stop valves are only fully opened or fully closed.

Intercept valves must close quickly in the event of a loss of load, and must be able to throttle steam into the turbine and assist in speed control when the speed is higher than rated. Some turbines feature combined reheat and stop valves which share a common seat and can travel through full stroke regardless of their relative positions. Adequate steam sealing is a necessity to prevent impurities in the steam from depositing in the stem-bushing clearances.

Chapter fourteen: Operations, maintenance and inspection services for steam systems

Reliable plant operation involves organized maintenance practices to ensure that equipment operates efficiently and economically. A good maintenance program is based on joint planning between maintenance, manufacturing, and operating personnel. Together they anticipate conditions which lead to lower-than-desired reliability and look for opportunities to improve upon repair schedules.

Major objectives of a plant maintenance program include continuity of plant operation, maintaining equipment to safety standards, managing spare-parts inventory and performing preventive maintenance. A well-organized and implemented maintenance program will help meet these objectives. By managing spare parts within a comprehensive preventive-maintenance program many problems can be averted. Of course, there will always be unscheduled, unexpected maintenance tasks to perform, but a well-planned preventive-maintenance program will minimize them.

An effective spare-parts inventory-control system is essential to a successful maintenance program. There must be sufficient inventory on hand or quickly available while minimizing inventory investment. Each item in inventory which is unused and not needed represents an economic waste. When this is multiplied by all the items which go unused for long periods as a result of poor control, it represents a significant amount of money.

More critical than idle inventory is production loss resulting from downtime caused by unavailable parts. In most plants, a shutdown caused by not having an essential spare part would result in a very significant loss of revenue.

Spare-parts inventory control will:
- Assure that there will be enough material on hand to accomplish the task.
- Identify obsolete parts which tie up capital and are costly to store and handle.
- Create a complete transaction-audit trail.
- Reveal what should be done now to improve cost effectiveness.

To establish a good preventive-maintenance system requires collecting maintenance records on each piece of equipment in the plant, establishing maintenance standards, and organizing a work-management program. Most large-equipment manufacturers provide a recommended schedule for maintenance to guarantee smooth, troublefree operation. While it is easy to follow a schedule for a large piece of equipment, the same attention should be given to smaller items in the plant.

Lines that carry steam and condensate, along with the valves that control flow and the steam traps used to separate the steam and release condensate, must be maintained. A first step in establishing a preventive-maintenance program for steam systems is to identify the component parts of the system, assess their condition, and bring these parts to a minimum level of repair. Steam and condensate control valves and steam traps can be identified by tag number. In addition, pipe must be inspected and steam leaks identified and qualified.

Steam leaks not only represent safety and operating hazards, they are also sources of lost energy. In a steam system with limited capacity these leaks may add up to the additional capacity needed to meet steam requirements. Of course, once repaired, reduced operating cost will result.

Repair of steam leaks should be done in a planned, orderly fashion. A survey of all steam leaks will define the magnitude of work to be done. This should be followed by estimating the approximate size of each leak to determine priority for repair.

Steam-leak survey

A systematic approach to locating and repairing steam leaks throughout the plant needs to be developed to ensure the success of a preventive-maintenance program. Determining the existing condition of the plant will enable the maintenance department to judge the magnitude of effort required to keep the plant maintained.

In carrying out the survey, sufficient information should be gathered for maintenance personnel to easily locate the leaks within the plant when the scheduled repairs are made. An assessment of when and under what conditions these repairs can be made needs to be made to help in the overall schedule of repairs.

While performing the survey, the use of steam-system schematics (distribution and piping diagrams) is a prerequisite to be sure that the entire system is covered. Upon detecting a leak its location should be recorded on a schematic. A tag with information about the leak and a work-order number is affixed as close to the leak area as possible. Information about the leak may include the size, type, and whether it can be isolated for repair while the plant is operating, or whether repair must be done during a shutdown period. Work-orders are used by maintenance personnel in the scheduling of repairs. Matching the work-order with the work-order number on the tag left at the leak helps verify that repairs are made. In subsequent surveys it can then be determined whether a leak has redeveloped after repair or whether repair had never been made.

Once all of the survey data have been collected they can be analyzed and compiled into a form that presents the important findings of the survey and recommended action to be taken. In a summary report the severity of steam leaks should be recognized, classifying them by type, line pressure, size, and how often they occur or re-occur. Categorizing leaks by type will also help personnel inventory the materials necessary for repair and prepare personnel for the number of leaks of each type that will be repaired.

Also helpful to the overall preventive-maintenance program is knowing how the number of steam leaks grow. A comparison is made of newly found leaks with those found in previous surveys. A list of recurring leaks is made showing how the past severity of leaks in previous surveys compares to the current severity of the leaks. In general it will be found that steam leaks, regardless of size, should be repaired as soon as possible.

Quantification of the size of the steam leak is a very important part of the overall effort to maintain the steam system. Visual techniques, which compare the size of a leak to other leaks of known sizes, work best. Standard leaks can be developed by drilling tapered holes into pipe caps to simulate orifices. The resulting plume of steam will be of a certain size. Using the equations for flow of steam through an orifice this standard leak can then be used to classify an existing leak. Judicious selection of standard leaks will enable the survey to be conducted so that all leaks can be quantified without having too many standards to compare leaks with.

Since many leaks do not produce a recognizable steam plume, due to superheat or obstructions, another method of classification must be used as a supplement. Sound is a good method. While not as accurate, it is an acceptable way of classifying some leaks.

Steam-trap survey

While the steam-leak survey identifies areas of energy loss, a steam-trap survey will identify those traps which do not operate efficiently. Again, the keynote to a well-run program is to organize the testing so that all traps are tested and evaluated for a repair/replace decision. Each steam trap should

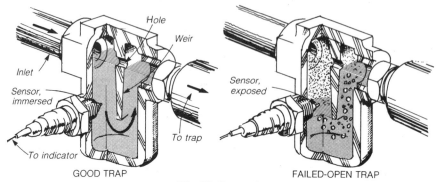

GOOD TRAP   FAILED-OPEN TRAP

Fig 18: Steam traps

have a number and a location associated with it (**Fig 18**).

Trap testing involves sound, sight, pressure, and/or temperature readings. Each of these is important in determining whether a trap is functioning properly. Quite often special training is given for testing of traps.

Before testing begins certain steam-system variables regarding the trap should be known:

- trap type
- location and application
- whether steam is saturated or superheated
- condensate load
- piping configuration
- whether pressure modulates or is steady
- amount of backpressure

Since it is not possible to see the trap operate, it is necessary to rely on one of several commonly used field-test methods.

Temperature and pressure measurements are a good starting point in determining the operability of a trap. Measuring the differential pressure across a trap will in many cases tell whether a trap is blowing steam or not. However, there are cases where this cannot be determined. For example, on a steam system having large return piping, a failed trap may not cause a rise in backpressure. The fact that there is no temperature differential across the trap means there is no indication of failure.

Temperature measurement can be crude or sophisticated. Touch will determine whether a trap is hot or cold. Actual measurement will give the exact temperature differential. Some more popular methods include the glove, squirt-gun, heat-sensitive crayon, pyrometer, and infrared tests. Glove and squirt-gun testing, the crudest tests, will tell whether the trap is hot or cold. This is done by feel in the case of glove testing, and by vaporization of water squirted into the trap from the gun. Heat-sensitive crayons are designed so that their material will change color at predetermined temperature levels. Pyrometers and infrared testers are more sensitive and naturally much more costly.

A more reliable method of testing, sound, requires a trained technician. He must be able to differentiate between a normally discharging trap and one that discharges live steam with its condensate. Here, operating principles and failure modes must be known. Testing equipment may include a screwdriver or stethoscope. The screwdriver's tip is pressed against the top of the trap while the tester holds the other end to his ear to hear performance. Stethoscopes are often linked to read-out gages, charts, and analog or digital indicators, many with printout capabilities.

When taking these types of readings, it is extremely important to filter out background noise. This is difficult in many cases since some traps may drain to a single header and sound can travel from trap to trap through piping. Several devices used for listening to traps have been designed to filter out all but specific frequencies. Some can even be used to test traps out of reach by use of extension poles.

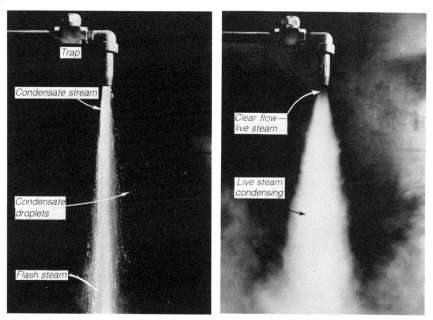

Fig 19: Blowing steam traps

The most reliable method of measurement is the visual one. Again, the tester must have substantial experience. On condensate-return lines, either a three-way valve or a pair of valves is needed to isolate the line and permit the trap to discharge to atmosphere. When the condensate hits the atmosphere it partially converts to flash steam. If live steam is escaping, its velocity will be much greater than that of flash steam, which appears as a slow-moving white cloud (**Fig 19**).

Another approach to visual examination is to vary the load on the trap. If an upstream valve is closed, it will result in a buildup of condensate in the line. Upon opening the valve a slug of condensate will hit the trap. The observer will hear or see this increased load. If live steam follows the condensate load it should be easily spotted.

In the case where an upstream bleedoff valve exists, reduced load can be placed on the trap. Again, an experienced tester should be able to monitor the trap for proper response to the reduced load.

Fig 20: Steam-trap testing

Each type of steam trap has a specific operating characteristic important to testing. Disc traps are usually tested with sound. In normal operation the trap will cycle because as the trap valve opens and closes it shuts against its seat, causing an audible sound. When the trap valve or seat becomes worn, it is possible to hear the steam leak. As the cycle rate increases, the possibility of wear or trap failure becomes greater. Operation of the trap can be heard better by pouring cold water on top of the trap (Fig 20).

For the two types of thermostatic traps, the bellows and the bimetallic, operating characteristics and testing procedures are the same. Continuous discharge is normal in the presence of condensate.   With progressive wear and leakage it becomes easier to detect trap failure. During a sound test a rush of condensate will be heard with the hiss of flash steam. As more live steam is present it becomes easier to determine that a trap has failed.

Float and thermostatic traps are modulating devices containing two valves. One of these valves is opened by a thermostatic element to vent air. The rushing sound of air and steam can be heard under normal operating conditions. While testing of the air-vent valve is not easy, it is very easy to tell when there is a failed-open vent valve because the venting is continuous and more intense than normal. Leakage in the condensate orifice occurs due to valve and seat erosion. Testing involves differences in sound attributable to condensate and live or flash steam. Visual testing is good if there is a gage glass. In this case, when the condensate level is down to the level of the orifice the trap will blow live steam. A minimum water level is required for proper float operation. Also, if the float is damaged it will malfunction resulting in a failed-closed trap.

The inverted-bucket trap discharges at full capacity and then shuts off. This on/off discharge can be both seen and heard. Since the bucket is suspended from the action lever, a distinctive rattle against the outer chamber by the bucket can be heard. Severe rattle and the sound of rushing steam point to loss of prime.

Air-vent pluggage will cause air and noncondensable gas to be trapped in the bucket, resulting in the trap failing closed.

Valve installation

After the proper valve has been specified care must be taken in planning its installation. To illustrate: Control-valve bodies and connections are usually durable and rugged. Even so, they are not designed to be force-fit into improperly aligned pipe. Improper pipe alignment results in line stresses, which must be relieved somewhere else in the pipe system. If these stresses are placed on the control valve it may result in misalignment of the stem, guide, and/or seat. For split-body valves this may cause the body flanges to separate. Hysteresis or leakage through the seat, flanges, or packing may be caused by this misalignment.

Installation of the control vlave should be such that the stem travels in a vertical plane with the diaphragm housing above the body. When a horizontal installation is desired, support braces may be necessary to support the diaphragm housing. If the valve is unsupported it may develop stem misalignment which could result in unacceptable hysteresis and packing leakage.

Valves should be located in easily accessible areas for adjustments and the periodic inspection required in the preventive-maintenance system. It may be desirable to utilize ladders and platforms to provide this access. When locating the valve it is necessary to determine whether the preferred location results in the valve being a collection point for rust, weld slag, and other objects. If it is unavoidable, temporary strainers or screens should be installed immediately upstream of the valve. A small piece of slag can ruin a well-lapped valve seat. For services which normally involve scale, dirt, or other solid foreign matter the installation of permanent strainers or filters is recommended (**Fig 21**).

Control-valve diaphragm housings are collectors of liquid and other foreign matter in the air system. Those valves which have positioners or booster relays attached are susceptible to failure when exposed to unclean-air sys-

Fig 21: Valve installation

tems. It is therefore imperative that a clean, dry, oil-free source of air be chosen for the control-valve operation.

During installation of a screwed valve the wrench should always be put on the hexagonal nearest the pipe. Supporting the valve in this manner prevents the applied torque from bending the valve. In supporting the valve suitable hangers placed close to both sides of the valve will reduce the pipe-transmitted stresses. For ball and gate valves, pipe should be threaded to the proper length so that the joint is tight before the pipe-end strikes the valve seat. Putting pipe compound on the pipe end rather than on the valve threads prevents the compound from getting on the valve seat where it may collect dirt and hinder tight shutoff.

Globe valves should be installed in the fully closed position. Otherwise, they can be twisted resulting in leakage due to the seatings not mating properly.

Control valves

After being in service for a short period of time control-valve packing will usually require readjustment and replenishment of the lubricating agent. If there are dirty or dusty conditions a plastic or rubber boot may be placed around the valve stem to protect the packing gland and guides in the bottom of the diaphragm housing. Other environmental conditions such as corrosive fumes or process leaks may indicate additional valve protection. An easy solution to this is to totally enclose the valve in plastic which does not hinder its performance. Installation accessories important to control-valve performance, such as tracing and lagging, also require post-startup inspection.

Once the initial post-startup maintenance has been completed a formal program should be initiated where valves are inspected and maintained at a regular frequency. Regular maintenance should include exercising and lubrication. Infrequently operated valves should be worked about once a month to ensure ease of operation. The greasing of spindles and gearing as well as other working components should be done about every six months. For the infrequently used valve that has not been operated for a long time, lubrication should be done prior to operation (Fig 22).

Stem packing should be inspected and replaced as necessary because some types of packing tend to harden, making movement of the valve element more difficult, especially when left inactive for long periods.

Fig 22: Control-valve inspection

Vibration can cause connecting bolts to loosen with time. These should be tightened. If there is also pipeline vibration the checking interval should be diminished.

When replacing renewable face or seat rings it is recommended that a rust solvent be employed. If the ring must be split or cut through to remove it, care must be taken to avoid damage to the body, particularly where there is a screwed ring. Graphite compounds used on the threads are very helpful in tightening screwed replacements. Inserting a shouldered ring should be done so the ring is tight against the shoulder.

For check valves it is important that they be regularly inspected to be sure that there is freedom of action and proper seating. Fatigued springs should be replaced.

Regular inspection and testing of emergency or automatic shutoff valves is mandatory. Under normal operating conditions these valves are not supposed to function, but they must function when the abnormal occurs.

Several items regarding gate valves and other valves with split rings need attention. When stuffing-box packing uses split rings these rings should be staggered so they are not all on the same side of the stem, forming a potential leakage source. Gland nuts should be uniformly tightened. When water seals are fitted they should be regularly inspected and the filling cup topped off. Regular scouring can be done to keep water free of grit—this helps save unnecessary wear and tear on valve faces and seats.

Safety valves

Since safety valves involve product liability, the manufacturers recommendations should be followed as closely as possible.

Safety valves are mounted with the spindle in the vertical position above the valve. To permit the withdrawal of internal components, installation should provide for sufficient clearance. Connections between the safety valve and boiler or superheater outlet should be as short as possible and usually of the same size as the valve.

Exhaust piping should be sized to accommodate full-valve capacity and should in no case be less than one size larger than the outlet branch. When exhaust piping is relatively long it should be fitted with an expansion joint to ensure that expansion and contraction movements are not transmitted to the valve body causing strain and possibly distortion and irregular operation of the valve. Vertical exhaust piping should be as straight as possible and anchored securely to resist the action of steam flow and to support the weight of the exhaust system.

Drain connections from the expansion-joint tee and valve body should be made independently to prevent steam from flowing back into the exhaust piping or body from other sources.

Once construction is complete and the system has been started up, the preventive-maintenance program calls for regular valve inspection to prevent unscheduled breakdowns.

Since safety and relief valves are for the protection of pressure vessels, boilers, superheaters, economizers, and pipe systems, they should be inspected regularly and maintained in top working order. Systematic and scheduled checking is the best method to ensure this.

Inspection of some valves is dictated by pressure-piping and pressure-vessel codes. Valve failure can be caused by improper installation, internal corrosion, and buildup of deposits.

A systematic, scheduled check of each valve in conjunction with a complete record-keeping system for tagged valves will help assure proper valve performance.

Depending on applicable codes and manufacturers' recommendations, some valves can be checked in place by lifting the disc off its seat while the valve is under normal operating pressure, then seeing if it reseats properly. If the valve does not reseat or if a damaged part is discovered, the valve should be replaced immediately.

Safety checks should be recorded on a safety-valve-inspection record sheet and kept in a central location. The test date should be mandatory rather than flexible to satisfy safety concerns and insurance and governmental parties. A good test frequency is one year since many plant turnarounds occur annually and the routine can thus be easily remembered.

Leakage of these valves at the operating pressure can be the result of damaged seats, tight lifting gear, distortion, or wrong operating pressure. Leakage or hissing may be an indication that one of these conditions exists. Remedial action can then be taken. Other inspection points include:

- Checking the spring assembly and spindle for corrosion, pitting, and cracking.
- Checking that the valve inlet is at a right angle relative to the outlet face. Piping strains may distort the body enough to cause internals to bind.
- Making sure that adjusting-ring pins are locked in place and sealed. Cap and lifting gear also should be installed and sealed.
- Making sure that tag data match written records and comparing them with boiler capacity and rating. Very often boilers or process systems are modified and the associated safety valves should be changed due to higher pressure rating or capacity.
- Removing all drain plugs.
- Checking that gags are not left on valves.
- Checking for residue buildup on valves.
- Checking that exhaust-piping muffler flow paths are unclogged.

For valves that have not operated since the previous inspection, manual operation is desirable to ensure the valve functions properly. In some cases, valves can be operated manually at design pressure by using a 20-foot rope to raise the lift lever to the full-open position. This manual operation can have harmful effects on boilers with trapped residue in the system. When steam forces the residue out of the system, the velocity with which it escapes, as well as particle size, may cause seat damage, resulting in leakage. It may be possible to prevent this by blowing steam through non-code valves prior to testing the safety valve. This will reduce the residue.

Valve records kept on visual testing should include nameplate data, date of last inspection and results, date of operation and results, date of testing and results, date of prior maintenance work and nature of repairs.

Valve repair

While a good preventive-maintenance program will limit the number of times a valve will fail, there will still be unplanned valve failures. The risk involved in not conducting a regular preventive-maintenance program is

Fig 23: Valve maintenance

having to make repairs under distress conditions. There may be loss of production or even plant shutdown. In either case, whether the valve is repaired during a scheduled or unscheduled period, maintenance and repair techniques for overhauling are the same (Fig 23, 24).

When a valve is sent out for repair it should be accompanied by a work-order containing as much information about the valve and its service as possible. When it arrives in the shop the actuator-connection orientation is marked in relation to body flanges. Once the components are all separated

Fig 24: Large disassembled valve

each one is inspected to determine the extent of repair necessary. When exposed surfaces are unfit for repainting they should be descaled. If sandblasting is used, masking tape can be used to protect all machined surfaces such as flange faces. Special greases can also protect surfaces if acid washes are used to remove them later.

Following descaling, the need for machine work is established. Flange surfaces should be inspected for damage such as scratches or grooves that will prevent proper gasket seating.

Screwed seat rings may also require replacement. Replacement is usually due to wire drawing, corrosion, or erosion. Machining will be sufficient to repair small pits and worn surfaces.

If replacement of threaded seat rings is necessary it should be completed before any machining is done to the body. This is because forces required to tighten the new seat ring may cause body distortion. Removal of threaded seat rings usually requires specially fitted tools which are usually provided by the valve manufacturer.

Seat rings often require a sealing compound to prevent leakage through the threaded area. Leakage will not only prevent tight shutoff but will also cause valve-body erosion and wear. The sealing material used should be that recommended by the manufacturer. After machining has been completed, painting of components is done. Masking will prevent paint from being applied to gasket surfaces.

Valve plugs, stems, and seat rings are precisely machined and often their replacement from the manufacturer is the only practical solution. The high polish on the control-valve stem is required to give the best possible packing seal. Plug and seat ring are fitted in concert, requiring precise machining and finishing.

Repair of valve actuators is also included in the valve maintenance. Each component should be carefully inspected after the actuator has been carefully disassembled. Because all valve stroking force is transmitted through the yoke, it should be inspected for signs of cracking. Component replacement is indicated when they have been either weakened or worn. The diaphragm should also be inspected for signs of cracking, hardening, or abrasion. Surface preparation and painting of actuator components is the same as for body components.

Threaded parts should be dressed for assembly and adjustment. Seals on reverse-acting diaphragm actuators should be replaced with new ones. Most important in reassembly is that gaskets of proper material and size be used and properly centered. Also of importance is the use of bolts of proper size and material. During bonnet bolting and blind-head flange bolting it is important that tightening be done in an even manner to assure even loading on the gasket. This is done by tightening diagonally opposite bolts evenly.

Packing-gland depth should be measured to determine the proper number of packing rings. Use of preformed, uniform packing rings assures that the best possible seal is achieved in the gland. For lantern glands a sufficient number of rings needs to be placed below the lantern ring to assure that it will mate with the lubrication port in the gland housing. The packing rings are inserted individually and lightly tapped into place because pounding of the rings can distort them. After the proper number of rings is placed in the gland, the follower is installed finger-tight for hysteresis testing of the assembled valve. Actuator reassembly also includes lubrication as specified by the supplier. The actuator stem, spring-thrust bearings and retainers, and the spring-adjusting nuts may all need lubrication.

APPENDIX: NOMOGRAPHS FOR STEAM DATA

Properties of saturated steam

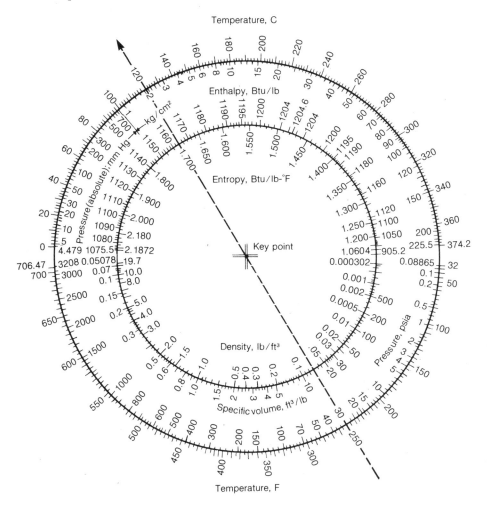

Directions: Line up one known quantity with the key point, draw the line, then read off the other desired values

Properties of superheated steam

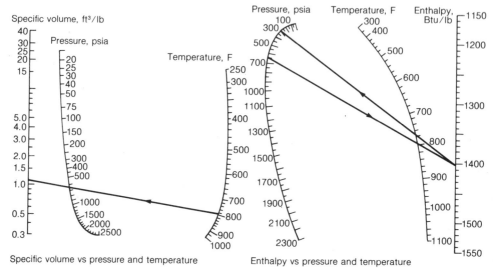

Specific volume, ft³/lb

Specific volume vs pressure and temperature

Enthalpy vs pressure and temperature

Directions: Draw a straight line between the pressure and temperature, then read the appropriate values of specific volume and/or enthalpy

Steam volume vs quality

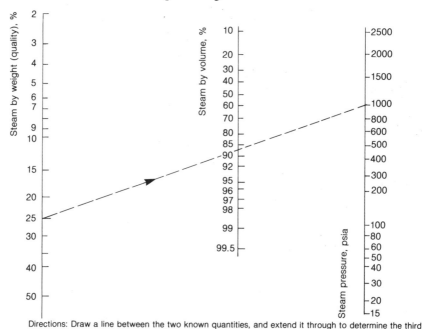

Directions: Draw a line between the two known quantities, and extend it through to determine the third

Steam flow through orifices

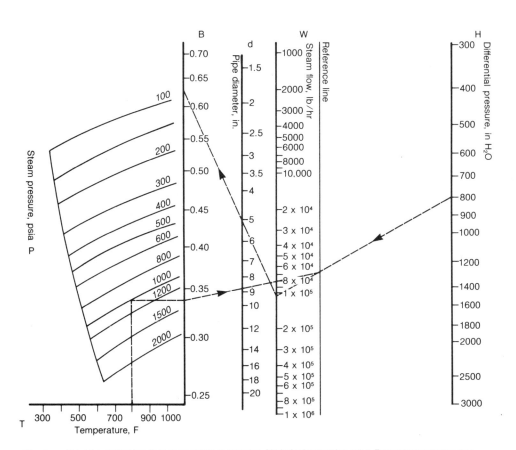

Directions: Extend a vertical line from temperature to pressure. Mark the intersection point. Extend horizontal line from this point to scale B. Mark intersection point. Connect this point with steam flow and extend line to reference line. Mark intersection point. Extend line from differential pressure to marked point on scale W. Connect point on scale W with pipe diameter. Extend to scale B and read value. Multiply this value by the pipe diameter to arrive at the orifice size.

Note: reverse the procedure to determine pressure drop across specific-sized orifices.

Pressure drop in steam lines

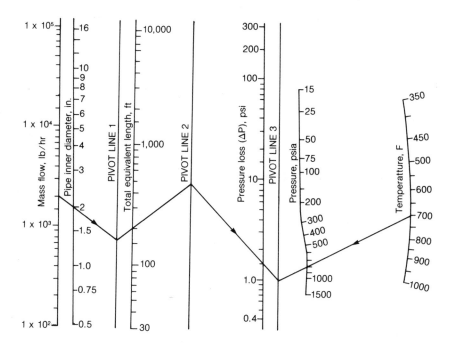

Directions: Knowing the flow, pipe diameter, steam temperature (T) and pressure (P), and equivalent pipe length, connect T and P, mark where the line intersects with pivot line 3 (C). Connect flow and pipe diameter, mark where the line intersects with pivot line 1(B). Connect B with the equivalent length of pipe value, and mark where it intersects pivot line 2(A). Connect (A) with (C), read where the line intersects with the pressure drop.

Condensate produced in bare
or insulated pipe

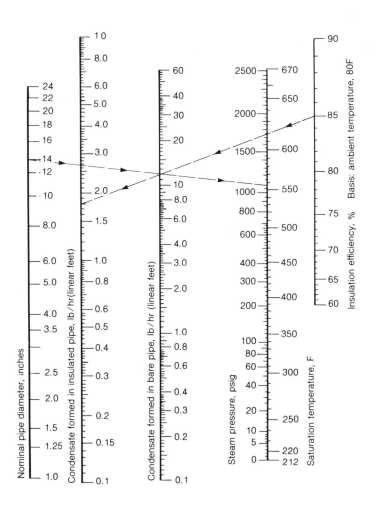

Directions: Connect pipe diameter with steam pressure (or temperature), read where the line crosses condensate-production in bare pipe scale. Then, align the insulation efficiency with the bare-pipe reading and note where it intersects the condensate produced in insulated pipe scale.

Amount of condensate that will flash to steam when pressure is lowered

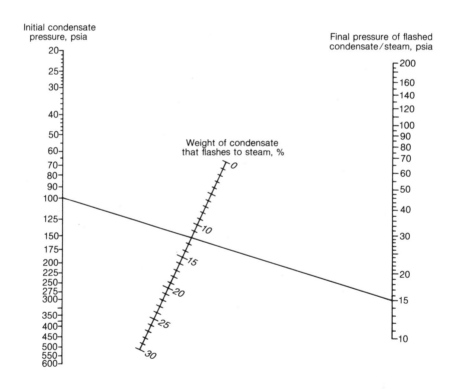

Directions: Align the initial and final pressure of condensate, then read the weight percent of flash steam

Steam requirements for air heating

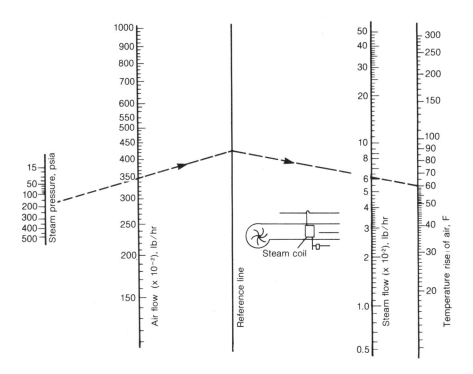

Directions: Connect steam pressure with flowrate of air, extend line to reference line. From this point, extend a line to the air temperature rise, then read where it intersects the steam flow scale.

Pressure rise due to water hammer

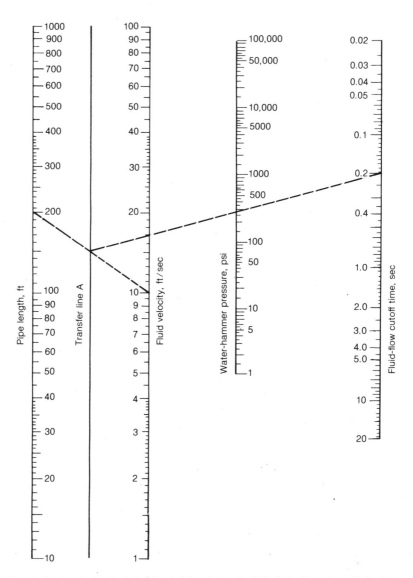

Directions: Extend a line from the length of pipe(ahead of the value) to the fluid velocity. From where this line intersects the transfer line A, draw a line to the cutoff time. Read where this line intersect's the water-hammer-pressure line.

Control-valve coefficients for steam (approximate)

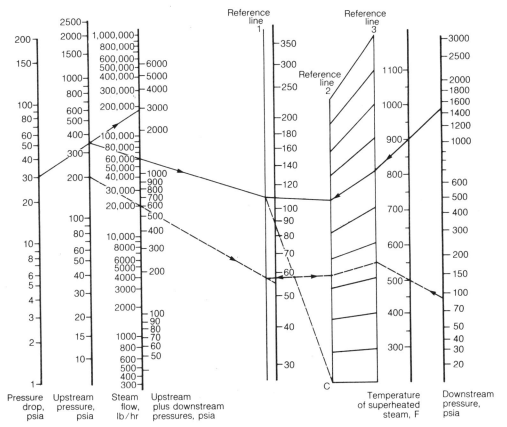

Directions: For non-critical flow(upstream pressure <2 x downstream press) align pressure drop with(upstream + downstream pressure). Align intersection of upstream-pressure scale with steam flow. Where this line intersects reference line 1 is A. Align downstream pressure with steam temperature, and extend it to intersect reference line 3, follow guidelines to reference line 2, this is point B. Align A and B and read where it crosses the control-valve-coefficient line(for superheated steam). For saturated steam, align A with C on reference line 2 and read off the control-valve-coefficient line.

For critical flow(upstream pressure >2 x downstream pressure), align upstream pressure with steam flow, to intersect reference line 1, called point D. Then align downstream pressure with steam temperature and extend it to intersect reference line 3. Follow guidelines to reference line 2, called point E. Align E and D and read control-valve coefficient for superheated steam.

REFERENCES

Section I: INDUSTRIAL STEAM SYSTEMS

1 J G Brennan and others, Food Engineering Operations, Elsevier Publishing Co Ltd, Amsterdam/London/New York, 1969

2 M Sittig, Pulp & Paper Manufacture: Energy Conservation and Pollution Prevention, Noyes Data Corp, Park Ridge, New Jersey, 1977

3 S G Cooper, The Textile Industry: Environmental Control and Energy Conservation, Noyes Data Corp, Park Ridge, New Jersey, 1978

4 M Sittig, Petroleum Refining Industry: Energy Saving and Environmental Control, Noyes Data Corp, Park Ridge, New Jersey, 1978

5 Industrial Energy Use, US Congress Office of Technology Assessment (OTA-E-198), Washington DC, June 1983

6 District Heating Handbook, fourth edition, International District Heating Association, Washington DC, 1983

7 ASHRAE Handbook, Systems Volume, The American Society of Heating, Refrigeration, and Air Conditioning Engineers, Inc, Atlanta, Georgia, 1980

8 J Reason, How electrode boilers can cut the cost of steam generation, Power May 1984, p 87

9 Solar projects marginally economic—even under favorable conditions, Power November 1981, p S. 14

10 R L Koral, editor, Foundations for the Solar Future, The Fairmont Press, Inc, Atlanta, 1981

11 M D Concannon, Condensate return systems, Plant Engineering, December 13, 1979, p 87

12 A J Haft and D E Nemeth, In-plant metering conserves energy, Power's Energy Systems Guidebook, 1978, p 95

13 W L Campagne, What's steam worth?, Hydrocarbon Processing, August 1981, p 117

14 W L Campagne, Convert steam balances into dollar balances, Hyrocarbon Processing, July 1983, p 57

15 W E Danekind, Steam management in a petroleum refinery, Chemical Engineering Progress, February 1979, p 51

16 Textile plants turn to central steam supply, Power's Energy Systems Guidebook, 1983, p 30

17 R G Schwieger, Industrial boilers: What's happening today, Power, February 1977, p S.1

18 M R Bary, Heat recovery from boiler blowdown—and its economics, Power's Energy Systems Guidebook, 1978, p 124

19 Making better use of your process steam supplies, Process Engineering, April 1983, p 43

20 R J Bender, Steam generation, Power, June 1964, p S. 1

21 Industrial boilers approach small utility units in capacity, pressure, Power, November 1981, p S. 23

22 A Watson, Blowdown-heat-recovery systems, Plant Engineering, June 28, 1984, p 53

23 L J Shapiro, Condensing turbine can improve economics of cogeneration, Power, August 1982, p 73

24 Gas turbine flexibility: Ideal for Purpa's demands, Power, January 1984, p 35

25 Turbine design emphasizes importance of electric power, Power, November 1983, p 166

26 D A Cline and K G Budden, Meeting industrial need for higher steam pressure, Power, March 1982, p 57

27 J Makansi, Advances in steam-turbine technology focus on efficiency, Power, July 1984, p 19

28 C J Kibert and W A Smith, Analysis of pressure reducing stations for cogeneration, Energy Engineering, Volume 80, No. 1, 1983 p 5

29 J Gilbert, Advances in piping and valve configurations for cogeneration systems, Specifying Engineer, July 1981, p 56

30 J Mower, Selecting condensate-handling systems, Plant Engineering, December 13, 1984, p 55

31 J Makansi and R G Schwieger, Fluidized-bed boilers, Power, August 1982, p S. 1

32 Steam Systems, Mobil Engineering Guide, Internal document of the Mobil Oil Corp, March 1978

33 M P Polsky, Scrutinize thermal-load data before you plan to cogenerate, *Power*, December 1983, p 29
34 Cogeneration: Full steam ahead, *Power's Energy System Guidebook*, 1978, p 115
35 B D Coffin, Compare total costs of cogeneration-system alternatives, *Power*, October 1984, p 59
36 Upgrading waste steam for process use, *Power*, March 1982, p 115
37 Steam: Its Generation and Use, The Babcock and Wilcox Co, New York, 1978
38 S P Gambhir, T J Heil, and T F Schuelke, Steam use and distribution, *Chemcial Engineering*, December 18, 1978, p 91

Section II: UTILITY STEAM SYSTEMS

1 R J Bender, Steam generation, *Power*, June 1964, p S. 1
2 Steam generation, *Power*, April 1982, p 161
3 R G Schwieger, Heat exchangers, *Power*, June 1970, p 34
4 O Martinez and J A Makuch, Variable-pressure operation both flexible and efficient, *Power*, January 1982, p 62
5 J F Riley, R L Schumacher and M R Bary, When sliding pressure operation can boost cycling-unit efficiency, *Power*, March 1984, p 45
6 Steam: Its Generation and Use, The Babcock and Wilcox Co, New York, 1978

Section III: STEAM-SYSTEM CONTROL THEORY

1 P Harriott, Process Control, McGraw-Hill, Inc, New York, 1964
2 C H Cho, Optimum boiler load allocation, *Instrumentation Technology*, October 1978
3 L L Fisher and R L Feeney, How important is turbine control?, *Chemical Engineering Progress*, March 1984
4 G R Fryling, Combustion Engineering, Inc, New York, 1967
5 H P Kallen, Handbook of Instrumentation and Controls, McGraw-Hill, Inc, New York, 1972
6 L Bengsson and others, Commercial experience with circulating fluidizing bed systems for cogeneration, Proceedings of the American Power Conference, 1981
7 Steam: Its generation and Use, The Babcock and Wilcox Co, New York, 1978
8 J Makansi and R G Schwieger, Fluidized-bed boilers, *Power*, August 1982, p S. 1
9 F D Gelineau, How to design boiler controls for waste fuels, *Power Energy Systems Guidebook*, 1982
10 W R Kelly and others, Industrial application of fluidized bed cogeneration systems, *Chemical Engineering Progress*, January 1984

Section IV: HARDWARE FOR STEAM SYSTEM CONTROL

1 R C King, editor, Piping Handbook, fifth edition, McGraw-Hill, Inc, New York, 1967, pp 20.1-20.50
2 D Coleman, Valve design evolving as utility conditions change, *Specifying Engineer*, July 1981, p 60
3 H Kawamura, Selecting valves for the HPI, *Hydrocarbon Processing*, August 1980, p 61
4 D K Sharma, Choose the correct safety-relief valve design to ensure optimal performance and useful life, *Power*, June 1981, p 66
5 S J Bailey, Control-valve prognosis 1980; analogs hold as digitals gain, *Control Engineering*, January 1980, p 64
6 J V Smyth, Understanding design and operation of multiturn gate, globe, angle, and diaphragm valves, *Plant Engineering*, July 10, 1980, p 79
7 Valves, *Power*, April 1982, pp 18-23
8 M McCoy, Understanding design and operation of relief and pressure-regulating valves, *Plant Engineering*, August 21, 1980, p 115

9 J Morgenroth, Understanding design and operation of quarter-turn plug, ball, and butterfly valves, *Plant Engineering,* July 24, 1980, p 64

10 L M Tierstein, Selecting temperature regulators and controllers, *Instrumentation Technology,* December 1979, p 52

11 W K White, Basic design features of diaphragm valves, *Plant Engineering,* February 17, 1983, p 56

12 G R Watts, Pressure regulators or control valves, *InTech,* June 1980, p 55

13 W O'Keefe, Check valves, *Power,* August 1976, p 25

14 W O'Keefe, Butterfly-valve improvements give designers new choices for modern piping systems, *Power,* September 1981, p 97

15 M Freeman, Designing nuclear control valves, *Power,* October 1974, p 39

16 B E Beakley, Understanding design and operation of balanced and unbalanced flow-control valves, *Plant Engineering,* January 7, 1982, p 71

17 W O'Keefe, Valves, *Power,* February 1983, p S. 1

18 W O'Keefe, Valve actuators, *Power,* April 1979, p S. 1

19 W O'Keefe, Instrument valves and accessories, *Power,* February 1982, p S. 1

20 W O'Keefe, Control valves, actuators, regulators, and positioners, *Power,* April 1976, p S. 1

21 S J Bailey, Process control valves meet energy challenge many ways, *Control Engineering,* March 1981, p 56

22 W O'Keefe, How valve manufacturers and users solve tomorrow's problems, *Power,* March 1984, p 61

23 T C Elliott, Key measurements in power and process, *Power,* September 1975, p S. 1

24 E R Cunningham, Positioning flow-control valves, *Plant Engineering,* November 12, 1981, p 82

25 W O'Keefe, Benefit from systems approach with today's safety valves, *Power,* January 1984, p 17

26 H D Baumann, Final control elements: Progress and prognostications, *InTech,* June 1979, p 62

27 G R McKillop, Come up to state-of-the-art on valve and pump packing, *Power,* June 1980, p 105

28 J Reason, Special-purpose flowmeters offer better accuracy, range, linearity, *Power,* March 1983, p 29

Section V: STEAM-SYSTEM CONTROL HARDWARE APPLICATIONS

1 P V Blarcom, Using desuperheaters for steam-temperature control, *Plant Engineering,* September 17, 1981, p 137

2 G A Keith, Steam-dump valve stability, (ISBN- 87664-364-0) Instrument Society of America, New York, 1977

3 W O'Keefe, Steam traps, *Power,* May 1984, p S. 1

4 D D Rosard and T McCloskey, Bypass systems increase cycling capability of drum boilers, *Power,* July 1984, p 81

5 D E A Gardner, How to apply existing bypass technology to drum boilers, *Power,* July 1984, p 95

6 G Masche, Systems Summary of a Westinghouse Pressurized-Water-Reactor Nuclear Powerplant, Westinghouse Electric Corp, 1971

7 G Dodero and N DiScioscio, Relate your control-valve and actuator applications to process characteristics, *Power,* July 1981, p 65

8 Proceedings of the Symposium on State-of-the-art Feedwater-Heater Technology, Electric Power Research Institute (report no. CS/NP-3743), Palo Alto, California, October 1984

9 W O'Keefe, Learn fluid-handling lessons from nuclear isolation valves and actuator systems, *Power,* January 1981, p 68

10 B G Skrotski, Steam turbines, *Power,* June 1962, pp S.24-S.31

11 R J Bell and E F Conley, Reliable feedwater heaters, presented at the ASME Joint Power Generation Conference, Toronto, September 1984

12 A H Klopfenstein, Control-valve applications for the power generation industry, Internal document of Fisher Controls International, Inc, Marshalltown, Iowa

13 D S Peiken, Advancements in control-valve technology, presented at the Fisher Controls International, Inc Power Seminar held in Des Moines, Iowa, October 1975

14 R A Uffer and S Farrington, Design of feedwater control valves, presented at the Instrument Society of America National Conference held in Philadelphia, 1978

15 L M Anderson, New extraction-control system for steam turbines, *Hydrocarbon Processing,* February 1978

16 W L Garvin, Keep boiler pressure safely below safety-valve set pressure, *Power,* December 1981, p 43

17 Desuperheating steam with water, Internal document of the Copes-Vulcan Div. of White Consolidated Industries, Lake City, Pennsylvania, 1983

18 The control of prime-mover speed: The controlled system, Internal document of the Woodward Governor Co, Rockford, Illinois

19 E K Kempers, How to cure desuperheater spray-valve headaches, *Power,* February 1984, p 35

20 Large steam turbines for reliable and efficient power generation, Internal document of The General Electric Co, Schenectady, New York

21 E B Woodruff, H B Lammers and T F Lammers, Steam-Plant Operation, fifth edition, McGraw-Hill Book Co, 1984

22 W O'Keefe, New ideas help feedpump recirculation valves meet severest pressure and flow needs, *Power,* March 1979, p 72

23 ASHRAE Handbook, Systems Volume, The American Society of Heating, Refrigeration, and Air Conditioning Engineers, Inc, Atlanta, 1980, pp 13.8-13.22

24 E R Cunningham, Solutions to valve operating problems, *Plant Engineering,* September 4, 1980

25 R S Perry, How to prevent valve problems, *Hydrocarbon Processing,* August 1980

26 W O'Keefe, What is the best way to close butterfly valves slowly, *Power,* May 1974, p 142

27 W O'Keefe, Valves, *Power,* February 1983, p S.1

28 W O'Keefe, Valve actuators, *Power,* April 1979, p S. 1

29 W O'Keefe, Instrument valves and accessories, *Power,* February 1982, p S.1

30 W O'Keefe, Check valves, *Power,* August 1976, p S.1

31 M Adams, Don't overspecify control valves, *Chemical Engineering,* October 29, 1984

32 J A Carey, Specifying tight control-valve shutoff can be costly, *Oil and Gas Journal,* November 23, 1981

33 D R Lazar and S T Terret, Steam-leak surveys as a technique to reduce energy losses, *Technical Association of the Pulp and Paper Industry (Tappi) Journal,* February 1983

34 J W Brock, Safety-valve maintenance and inspection give you ensured boiler pressure relief, *Power,* August, 1980

35 D A Keech, Sharpen steam-trap testing skills, *Power Energy Systems Guidebook,* 1982

36 T W Riemer, A couple of suggestions for energy conservation, *Tappi Journal,* February 1983

INDEX

Artwork Credits

SECTION 1:
INDUSTRIAL STEAM SYSTEMS

1. Steam systems, Mobil Engineering Guide, Internal document of the Mobil Oil Corp, March 1978

2. J G Brennan and others, Food Engineering Operations, Elsevier Publishing Co, Ltd, Amsterdam / London / New York, 1969, p 230

3. Same as above, p 210

4. Energy Engineering, Volume 80, No 1 1983, p 5

5. Power, June 1964, pp S.2-3

6. Same as above, p S.12

7. Same as above, p S.12

8. Same as above, pp S.18-19

9. Power, February 1977, p S.7

10. Power, February 1978, p S. 18

11. Power, May 1984, p 89

12. Power, Energy Systems Guidebook, 1982, p 18

13. Power, August 1982, p S.7

14. Power, August 1982, p S.6

15. Power, Feb 1978, p S.6

16. R L Koral, editor, Foundations for the Solar Future, The Fairmont Press Inc, Atlanta, 1981, p 187

17. Same as above

18. Chemical Engineering, December 1978, p 92

19. Steam Systems, Mobile Engineering Guide, Internal document of the Mobil Oil Corp, March 1978

20. Same as above

21. Plant Engineering, December 13, 1979, p 87

22. Same as 19

23. Same as 21

24. Power, December 1983, p 30

25. Power, Energy Systems Guidebook, 1978, p 115

26. Power, October 1984, p 59

27. Power, March 1982, p 115

SECTION 2:
UTILITY STEAM SYSTEMS

1. Power, June 1964, p S.9

2. Same as above

3. Power, Special Report, Power Generation Technology, April 1982, p 63

4. Same as above, p 63

5. Same as above, p 65

6. Same as above, p 64

7. Same as above, p 66

8. Same as above, p 68

9. Power, April 1983, p 6

10. Courtesy of the Brown Boveri Corp, North Brunswick, New Jersey

11. Power, June 1962, p S.18

12. Same as 3, p 144

SECTION 3:
STEAM-SYSTEM CONTROL THEORY

1,2,3,4,5,6,7,8,9. None

10. Steam: Its generation and use, The Babcock & Wilcox Co, New York, 1972, p 35-7

11. Same as above, p 35-14

12. Same as above, p 35-15

13. Same as above, p 35-15

14. Same as above, p 35-23

15. Same as above, p 35-15

SECTION 4:
HARDWARE FOR
STEAM-SYSTEM CONTROL

All photographs and artwork courtesy of POWER Magazine and LESLIE Co.

SECTION 5:
STEAM-SYSTEM CONTROL
HARDWARE APPLICATIONS

1. LESLIE Co

2. R A Uffer and S Farrington, Design of feedwater control valves, presented at the Instrument Society of America's National Conference held in Philadelphia, 1978

3. Internal document of Masoneilan International, Inc, Norwood, Mass

4. LESLIE Co

5. Power, February 1984, p 36

6. Power, Energy Systems Guidebook, p 124

7. Electric Power Research Institute (EPRI), Palo Alto, California, Report No CS/NP-3743, October 1984, p 4-103

8. Internal document of Masoneilan International, Inc, Norwood, Mass

9. Power, July 1981, p 65

10. Power, July 1984, p 96

11. G A Kieth, Steam-dump valve stability (ISBN-87664-364-0), Instrument Society of America, New York, 1977

12. LESLIE Co

13. LESLIE Co

14. Power, May 1984, p S.13

15. Power, May 1984, p S.11

16. Steam-turbine generators, Internal document of Turbodyne Corp, Wellsville, New York

17. Large steam turbines for reliable and efficient power generation, Internal document of the General Electric Co, Schenectady, New York

18. Power, May 1984, p S.16

19. Same as above, p S.16

20. Same as above, p S.15

21. None

22. None

23. Power, February 1983, p S.48

24. None

Nomograph sources

1. Properties of saturated steam, Heating-/Piping/Air Conditioning Journal, June 1979, p 393

2. Superheated-steam properties at a glance, excerpted by special permission from Chemical Engineering, January 9, 1985, Copyright 1985, by McGraw-Hill, Inc., New York, N.Y. 10020

3. Nomogram converts between steam volume and weight, Oil and Gas Journal, January 22, 1979, p 74

4. Chart gives pressure loss in steam lines, Oil and Gas Journal, January 9, 1978, p 95

5. Estimating condensate produced in bare or insulated pipe, Plant Engineering, February 7, 1980, p 78

6. How much condensate will flash?, excerpted by special permission from Chemical Engineering, May 14, 1984, copyright 1984, by McGraw-Hill, Inc., New York, N.Y. 10020

7. Nomograph calculates steam requirements for air heating, Heating/Piping/Air Conditioning Journal. September 1978

8. Nomograph determines pressure rise due to water hammer, Heating/Piping/Air Conditioning Journal, June 1979, p 405

9. Control valve coefficients for steam, Plant Engineering, August 20, 1981, p 80

10. Estimating steam flow through orifices, Plant Engineering, September 27, 1984, p 48